EUI — Series A — 2.5
Cappelletti/Seccombe/Weiler (Gen. Eds.)
Integration Through Law
Vol. 5: Daintith/Williams, The Legal Integration of Energy Markets

European University Institute
Institut Universitaire Européen
Europäisches Hochschulinstitut
Istituto Universitario Europeo

Series A

Law/Droit/Recht/Diritto

2.5

Badia Fiesolana — Firenze

Integration Through Law

Europe and the American Federal Experience

A Series under the General Editorship of

Mauro Cappelletti · Monica Seccombe · Joseph Weiler

Volume 5

The Legal Integration of Energy Markets

by

Terence Daintith and Stephen F. Williams

1987

Walter de Gruyter · Berlin · New York

Library of Congress Cataloging-in-Publication Data
(Revised for volume 5)

Integration through law.
(Series A, Law / European University Institute = Series A, Droit / Institut Universitaire Européen ; 2.1.-)
Includes bibliographical references and indexes.
Contents: v. 1. Methods, tools, and institutions. bk. 1. A political, legal, and economic overview. bk. 2. Political organs, integration techniques, and judicial process. bk. 3. Forces and potential for a European identity -- v. 2. Environmental protection policy / by Eckard Rehbinder and Richard Stewart. -- v. 5. The legal integration of energy markets / by Terence Daintith and Stephen F. Williams.
1. Law--European Economic Community countries.
2. Law--United States. 3. Federal government.
I. Cappelletti, Mauro. II. Seccombe, Monica. III. Weiler, Joseph, 1951- . IV. Series: Series A--Law ; 2.2, etc.
KJE969.I58 1985 340'.2 85-16320
342
ISBN 0-89925-112-9 (U.S. : v. 2)

CIP-Kurztitelaufnahme der Deutschen Bibliothek

Integration through law : Europe and the American federal experience / a ser. under the general editorship of Mauro Cappelletti . . . – Berlin ; New York : de Gruyter
(European University Institute : Ser. A, Law ; 2)
NE: Cappelletti, Mauro [Hrsg.]; Istituto Universitario Europeo (Fiesole): European University Institute / A
Vol. 5. Daintith, Terence: The legal integration of energy markets. – 1987

Daintith, Terence:
The legal integration of energy markets / by Terence Daintith and Stephen F. Williams. – Berlin ; New York : de Gruyter, 1987.
(Integration through law ; Vol. 5)
(European University Institute : Ser. A, Law ; 2)
ISBN 3-11-010743-0
NE: Williams, Stephen F.:

Dust Cover Design: Rudolf Hübler, Berlin. — Setting: Satz-Rechen-Zentrum, Berlin.
Printing: H. Heenemann GmbH & Co., Berlin.
Binding: Verlagsbuchbinderei Dieter Mikolai, Berlin.
Printed in Germany.

General Editors' Foreword

The Florence Integration Through Law Series is the product of a research project centered in the Law Department of the European University Institute, and as such it reflects the research interests of the Department: it is a contextual examination of European legal developments in comparative perspective. In the general introduction to the Series (published in Book One of Volume I), we explained fully the philosophy, methodology and scope of the Project. Here we wish merely to recapitulate some of the principal themes of special relevance to this Volume on Energy Policy.

The European Legal Integration Project set out to examine the role of law in, and the legal impact of, integration in Europe, using the United States federal system as a comparative point of reference. The Project was conceived and executed in two parts. In Part One (published in Volume I entitled "Methods, Tools and Institutions"[1]) a number of teams of American and European scholars examined a wide range of legal techniques and mechanisms for integration and undertook an overall general analysis of law and integration. The first book of Volume I ("A Political, Legal, and Economic Overview") establishes the comparative and interdisciplinary context, providing background studies on the political, legal and economic implications of integration in Europe and America and including studies on other federal systems (Australia, Canada, Germany and Switzerland) to add comparative perspective. The second book ("Political Organs, Integration Techniques, and Judicial Process") analyses the pre- and post-normative stages of the integration process, examining the decision-making and implementation problems, and the role of political and judicial organs therein, and describing the various forms of normative techniques available in a federal or supranational context.

The third and final book of Volume I ("Forces and Potential for a European Identity") focusses on how the law can be harnessed to promote the governmental or integrational objectives of union. It isolates for consideration some substantive goals (foreign policy, free movement of goods and persons, human rights protection and legal education), in order to elucidate the ways in which law has been or can be used to promote substantive objectives. This approach is more fully developed in the studies in Part Two of the Project which deals in greater detail with substantive areas of federal/transnational policy and is open-ended. To date, in addition to the present volume on energy policy, monographs have been published in the areas of environmental policy[2] and

[1] Published by Walter de Gruyter (Berlin/New York) in January 1986.

[2] E. Rehbinder & R. Stewart, Environmental Protection Policy (Vol. 2 Integration Through Law) was published by Walter de Gruyter (Berlin/New York) in December 1985.

consumer protection policy,[3] and others are planned in the areas of harmonization of corporation law and capital markets, as well as regional policy.

Why should we have chosen to include energy policy among the few substantive areas on which the Project would focus? Energy policy has a strange history in the Community. It seemed extremely important at inception — for political as well as economic reasons — an importance which found its expression in the Coal and Steel Community. But as the European states moved rapidly towards a cheap oil economy the Coal and Steel provisions lost much of their symbolic and economic significance. Euratom, which is also in some ways a Treaty of vision, was conceived in such a way that it had little practical effect. Thus during the 1960s, the period of major strides in "negative integration" and the "Heroic Period" of European legal integration, energy policy was overshadowed. But in the 1970s and 1980s the situation changed, for obvious reasons. The 1973 oil embargo — which represented, *inter alia*, a major challenge to the economic and political cohesion of the Community — was the first cause of change, or at least of a need for change. The emergence of Britain as a major oil producer was another such cause. And then there was the rapid cycle of shift in supply and demand which strained attempts to develop coherent policies.

Energy policy thus constitutes a fascinating case study of Community activity. In substance, if not in form, it is a second generation policy, for it is a policy which represents the operation of the "positive state" in the mixed economy. It is far more representative of current exigencies than the first generation policies of the "negative" kind which simply required the Member States to eliminate certain practices.

Energy is of course a vital policy area central to the economic well-being of the polity. At the same time it raises in most acute form important "federalism" problems. To what extent, if at all, should the Community, rather than the Member States individually, attempt to regulate the energy market? And if there is to be transnational integration, is the Community necessarily the best forum? To what extent can the Community control such policy within the political constraints currently in operation? These federal issues apart, there is the additional question of the extent to which regulation, per se, is altogether desirable.

Energy policy as a topic for legal analysis has, of course, already elicited a substantial amount of scholarly attention. Why then present this new examination? First and foremost is the value of a fresh analysis by the distinguished authors of this volume. But in addition, it is our belief that the Integration Project provided a special context for specific and unique insights. Most obvious is the comparative context: one of the underlying objectives of the studies in Part Two of the Project was to see what, if any, lessons the European Com-

[3] T. Bourgoignie & D. Trubek (with L. Trubek & D. Stingl), Consumer Law, Common Markets and Federalism in Europe and the United States (Vol. 3. Integration Through Law) was published by Walter de Gruyter (Berlin/New York) in 1986.

munity could learn from the United States in selected areas of substantive legal integration (and to some extent vice versa), and in fulfilment of this mandate the authors of this study present a tight comparative analysis of the European and American experiences. Through this comparative analysis we gain a better understanding of the problems associated with, for example, crossfrontier energy issues; we also gain a better understanding of the workings of the transnational/federal system of governance. There is indeed much to be learnt from this volume on the legal dimensions of energy policy.

The Project has invited, however, more than the comparative contribution. The Florence Integration Through Law Series is dedicated to the concept of Law in Context: the examination of legal problems in their political, economic and social setting. There has been much pontification in recent years about the value of interdisciplinarity. Implementation of this value, however, often falls short of much hallowed theoretical expectations. In this regard our claims were modest; we did not ask our contributors to bring the full scientific paraphernalia of, say, economics or political science to bear on their subject. We simply asked that the legal analysis be situated in, and be sensitive to, the implications of the socio-economic and political context. The present volume is, in our view, an extraordinarily successful example of this approach.

The European Integration Project follows on from an earlier wide-ranging research project which was carried out at the European University Institute — the Florence Access-to-Justice Project. Access to Justice was not only concerned with an examination and, indeed, extension of the procedural and institutional mechanisms for the vindication of rights in contemporary society. It was an approach which sought to emphasise that, in legal study, an analysis of the normative content of legal rules and policies — while still central — can give only a partial picture of the function and shortcomings of the law in its societal context. Normative analysis is but one layer of analysis: the effective (or otherwise) reach of the law, its implementation and enforcement, its accessibility to subjects to whom it is addressed as a source of rights and duties, is a second no less important layer. This approach has been a constant guideline to all contributors to the European Integration Project.

If the Access-to-Justice philosophy postulated the addition of this post-normative layer in the analysis of law, the institutional and procedural character of the Integration Project postulated the addition of yet another layer — a pre-normative layer. Both in the first general methodological part of the Project and in its second substantive part we have given considerable attention to the decision-making process by and through which norms emerge. The necessity of this addition is so clear as to obviate any lengthy explanation. Not only is decision-making an essential component in the analysis of the system as a whole, but it also gives, particularly in the context of the European transnational concordance of interests, an insight into the normative outcome and, as explained throughout the Project, into the very problems of implementation, application and enforcement. This study on energy policy, as well as other studies in Part Two of the Project, has adopted what one may call a "total" approach to legal analysis. Certainly the normative, "black letter" dimension of the law is explored; but this normative analysis is sandwiched between the pre- and post-normative phases. The present volume explores fully the process of policy-making, the difficulty it encounters and the political context against which normative compromises are reached.

The Integration Through Law Series represents a collective effort over a long period of time. At its inception we believed that the first methodological part of the Project would be the setting against which the subsequent substantive parts, such as this study on energy policy, would be written. Things often do not turn out as they were planned. The two parts of the Project in fact evolved simultaneously, and while the Part Two studies undoubtedly did rely on the general methodological background studies of Part One, the studies in Part One equally drew upon the analysis contained in the concrete substantive studies of Part Two. The work of Professor Daintith and Professor (now Judge) Williams assumed a special importance in the context of the Project. It brought economic analysis to the fore and also introduced a healthy scepticism about automatic assumptions concerning the value of integration. We thank them for their stimulating and authoritative contribution to the common effort.

Florence, January 1986

Mauro Cappelletti
Monica Seccombe
Joseph Weiler

Authors' Preface

As the series editors indicate in their Foreword to this volume, our mandate in this study was to see what lessons the European Community could learn from the United States in an area of substantive legal integration: that of energy. This was a daunting task: at first sight the differences in this sector between the Community and the United States appeared to be so profound — in terms of energy import dependence, energy market structures, and the degree of integration attained — as to preclude all useful comparison. Yet in the course of a lengthy process of investigation and discussion, with the stimulus of the December 1981 colloquium which united all participants in the Project, and with the encouragement of the series editors, we became convinced that certain aspects of energy policy in the United States and the Community could be profitably compared by reference to their common basic structures for legal integration: the removal of barriers to internal trade, and the creation of central decision-taking capacity. In order to focus effectively on the operation of these fundamentals of legal integration in the energy field, we have adopted a highly selective approach, concentrating our attention on the energy resources which have proved most problematical from the point of view of the construction and maintenance of a unified internal market. This means that for the most part we talk about oil and gas, with only passing references to coal and uranium. We have not set out to provide a general comparison of United States and Community energy policies, or even of the legal frameworks and implementation of such policies: our objective has rather been to examine and evaluate the legal structures for creating and maintaining unified energy markets in the United States and the Community, and to say what reciprocal lessons are offered by such evaluation.

We should like jointly to acknowledge the help and support we have received from the series editors. Separately, Professor Daintith wishes to thank Tony Curran and Leigh Hancher for indefatigable research assistance and text editing, and Anne-Lise Strahtmann for the typing of large numbers of drafts; Judge Williams, for his part, would like to thank Greg Berger for his research and Anne Guthrie for typing.

San Domenico di Fiesole
Boulder, Colorado
January 1986

Terence Daintith *
Stephen Williams **

* Professor of Law, European University Institute; Honorary Visiting Professor, Centre for Petroleum and Mineral Law Studies, University of Dundee.

** Judge, United States Court of Appeals for the District of Columbia Circuit; formerly Professor of Law, University of Colorado.

Acknowledgements

This is one of several volumes on "Integration Through Law: Europe and the American Federal Experience" being published under the joint sponsorship of the European University Institute and the Ford Foundation, which together funded an international research project on "Methods, Tools and Potential for European Legal Integration in Light of the American Federal Experience" co-directed by Professors Mauro Cappelletti and Joseph Weiler. The Project, headquartered at the European University Institute in Florence, involved the participation of over forty scholars mostly from Europe and America. In addition to thanking the European University Institute and the Ford Foundation which were the principal sponsors, the Project Directors would like to express their gratitude to the Institutions of the European Communities, particularly the Commission, whose contribution has been most encouraging. A special acknowledgement is due to Professor Martin Shapiro of the University of California at Berkeley, who coordinated the American contributions to the Project.

Table of Contents

List of Tables

Table of Cases

European Court of Justice

International Court of Justice

Table of Constitutional and Treaty Provisions

Law of the Sea Conventions

Table of Legislative Provisions

European Communities

Regulations

Decisions (ECSC)

Decisions (EEC, Euratom)

Directives

List of Abbreviations

ABF Research J.	American Bar Foundation Research Journal
ADA	Americans for Democratic Action
A.J.I.L.	American Journal of International Law
All ER	All England Law Reports
Am. J. Comp. L.	American Journal of Comparative Law
ATIC	Association Technique de l'Importation du Charbon
BGC	British Gas Corporation
BNFL	British Nuclear Fuels Limited
BNOC	British National Oil Corporation
BTU	British Thermal Unit
Bull. EC	Bulletin of the European Communities
CDF	Charbonnages de France
C.E.	Conseil d'Etat
CEA	Commissariat pour l'Energie Atomique
CECA	Communauté européenne du charbon et de l'acier
CEGB	Central Electricity Generating Board
CFP	Compagnie Française des Pétroles
C.F.R.	Code of Federal Regulations
Cmd.	Command Paper (HMSO, London)
C.M.L.R.	Common Market Law Reports
C.M.L. Rev.	Common Market Law Review
CNR	Compagnie Nationale du Rhône
Cogema	Compagnie Générale des Matières Nucléaires
Cong. (Rec.)	Congress(ional Record)
DOE	Department of Energy
Doc. COM	Commission Documents (EC)
Doc. SEC	Documents Secrétariat (EC Commission)
DONG	Dansk Olie og naturgas A/S
EAEC	*see* Euratom
EC	European Community, European Communities
E.C.R.	European Court (of Justice), Reports of the
ECSC	European Coal & Steel Community
EDF	Electricité de France
EEC	European Economic Community
EIA	Energy Information Administration
ENEL	Ente Nazionale per l'Energia Elettrica
ENI	Ente Nazionale Idrocarburi
Env. L. Rev.	Environmental Law Review
Env't Rep. Cas. (BNA)	Environment Reporter Cases (Bureau of National Affairs, Inc., Washington D.C.)
EPA	Environmental Protection Agency
EPAA	Emergency Petroleum Allocation Act

EPCA	Energy Policy & Conservation Act
ESA	Euratom Supply Agency
Euratom	European Atomic Energy Community
Eur. L. Rev.	European Law Review
F.2d	Federal Reporter, Second Series
Fed. Reg.	Federal Register
Fed. Supp.	Federal Supplement
FERC	Federal Energy Regulatory Commission
FPC	Federal Power Commission
FTC	Federal Trade Commission
GDF	Gaz de France
GNP	Gross National Product
Harv. L. Rev.	Harvard Law Review
H.R. Rep.	House of Representatives Report
I.C.J.	International Court of Justice (Reports)
IEA	International Energy Agency
IEP	International Energy Programme
I.L.M.	International Legal Materials
IRI	Istituto per la Ricostruzione Industriale
J. Energy & Nat. Res. L.	Journal of Energy & Natural Resources Law
J. Law & Econ.	Journal of Law & Economics
J. Legal Stud.	Journal of Legal Studies
J.O.	Journal Officiel (of EC unless otherwise specified)
Lloyd's Rep.	Lloyd's List Reports
Mich. L. Rev.	Michigan Law Review
MTOE	Million Tonnes of Oil Equivalent
NAM	Nationale Aardolie Maatschappij
Nat. Res. J.	Natural Resources Journal
Nat. Res. Law.	Natural Resources Lawyer
Nat. Tax. J.	National Taxation Journal
NCB	National Coal Board
N.E.2d	Northeastern Reporter
NGPA	Natural Gas Policy Act
OECD	Organisation for Economic Cooperation & Development
OEEC	Organisation of European Economic Cooperation
Ohio St. 2d	Ohio State Reports
O.J.	Official Journal (of EC unless otherwise specified)
OPEC	Organisation of Petroleum Exporting Countries
Ore. L. Rev.	Oregon Law Review
P.L.	Public Law
PWR	Pressurised Water Reactor
Rev. Dr. Pub.	Revue de droit public
R.M.C.	Revue du Marché Commun
RWE	Rheinisch Westfälische Elektrizitätswerke

S.I.	Statutory Instruments
S. Ct.	Supreme Court (Reporter)
S. Ct. Econ. Rev.	Supreme Court Economic Review
S.E.W.	Sociaal-Economische Wetgeving — Tijdschrift voor Europees en economisch recht
SNEA	Société Nationale Elf-Aquitaine
SNGSO	Société Nationale du Gaz du Sud-Ouest
So.2d	Southern Reporter
Stat.	US Statutes at Large
Stb.	Staatsblad voor het Koninkrijk der Nederlanden
Supt. of Docs.	Superindendent of Documents
Tex. L. Rev.	Texas Law Review
Tulane L. Rev.	Tulane Law Review
U. Chi. L. Rev.	University of Chicago Law Review
U. Colo. L. Rev.	University of Colorado Law Review
U.S.	United States Supreme Court Reports
U.S.C.	United States Code
U.S. Code Cong. & Ad. News	US Code Congressional & Administrative News
U.N.T.S.	United States Treaty Series
Utah L. Rev.	Utah Law Review
Va. L. Rev.	Virginia Law Review
VEW	Vereinigte Elektrizitätswerke Westfalen
W.L.R.	Weekly Law Reports (UK)
Yale L. J.	Yale Law Journal

Note on Numerical Terminology

Billion = 1,000 million
Trillion = 1,000 billion = 1 million million
Quadrillion = 1,000 trillion = 1,000 million million

Short ton = 2,000 lbs = c. 907 kg
Ton = long ton = 2,240 lbs = c. 1,016 kg
Tonne = metric ton = 1,000 kg (35 lbs less than a long ton)

Chapter One

Introduction

I. Energy Resources as an Integration Problem

Our research has been addressed to tension between integrative and disintegrative forces in the area of energy-related exhaustible natural resources: coal, oil, natural gas, and nuclear source materials. With occasional exceptions, we have not considered the electricity-generating phase of the energy process. Nor have we considered other natural resources, save insofar as their treatment by the law illuminates our argument on energy-related resources (as is the case, for the European Communities, with fish), or where they are critical to the production of energy-related resources (as is the case with water in the United States).

Energy-related exhaustible natural resources have certain characteristics that, when taken together, make them an exceptional point of stress for any federal or quasi-federal system. Here we concentrate on three of these: the local sense of possession of these resources (A); the problems of defining property rights in some energy-resources (B); and the importance of the regime for energy to the operation of a unified market (C).

A. The Local Sense of Possession

Local communities have displayed in recent years a strong sense of possessiveness towards depletable natural resources. This sense has been expressed in a variety of forms and contexts, from Robert Lafont's critique of the exploitation of the natural gas of Lacq, in France, as "*colonialisme intérieur*",[1] to the Scottish National Party's 1974 election slogan, "It's Scotland's Oil!" Use of such possessive forms is not, of course, intended to reflect the legal situation. Most of the time, in the United States, title to the resource is in private citizens (who may or may not be local citizens), private corporations (whose owners may be local only to a very slight degree), or even the federal government (as, for example, with most Montana coal). The same disjunction between posses-

[1] R. Lafont, La Révolution Régionaliste 154-58 (1967).

sive language and legal title appears in Europe where, although resources are normally owned by the State or by state public corporations, such language is frequently used, as in the examples above, by or on behalf of populations of sub-national localities.

Use of phrases such as "our" coal, "our" gas, and so on, should rather be seen as representing a moral and political claim based on social and economic concerns. Some adverse impacts of resource development are very local indeed. Most obvious are physical externalities, such as air, water, and visual pollution. Social side effects of the boom-and-bust cycle that may accompany development of an exhaustible resource may be more serious. The boom phase brings an influx of temporary workers, the displacement of settled local personal relationships and land-use patterns, a strain upon public facilities and on social services, and high crime and divorce rates. If resource development then declines, however, the local community may be left with only the derelicts of all this activity (plus, perhaps, a legal obligation to pay off debts incurred to handle the now-departed multitudes). Moreover its prior economic base may have weakened: traditional local industries may have ceased to be viable by reason of the bidding-up of wage rates in the period of boom, and may find it impossible to re-establish themselves when it is over.[2]

Even where the boom-and-bust cycle is not a serious peril, the depletable character of a natural resource may contribute to possessive local attitudes. As people are inclined to think of a depletable resource largely in physical terms, they will often be anxious about the time when depletion will cut off its contribution to local prosperity. (Often, of course, the anxiety may prove justified. In other instances, however, developments that reduce extraction costs, or increase demand, may give extraction of a given resource a life-span far exceeding that of many non-extractive enterprises.)

The local sense of possession may be fed not only by a perception of threat in natural resource development but also by a sense of opportunity. Nature in her perversity has distributed depletable natural resources most unevenly over the globe and within the two unions considered in this Project. Local control of production, if it can be achieved, may not only make it possible to palliate adverse local impacts but may also permit local populations to extract a return from production which may be applied to increase local wealth, prime new industrial investment, and so on. When we turn our attention to energy-related depletable resources, the dramatic price increases over the period 1973-1981 also come into play. Such increases make it obvious that for many reserves — and not only of petroleum — a significant proportion of the value at the wellhead or minehead may be "economic rent" (*i.e.*, in excess of the price needed to induce extraction). Any reserve for which that is true is an inviting target

[2] An example is the difficulties experienced by the traditional knitting and fishing industries of the Shetland Islands in Scotland, now a major base and terminal for North Sea oil operations: *see* T.M. Lewis & I.H. McNicoll, North Sea Oil and Scotland's Economic Prospects 122-25 (1978).

for local (and other) legislators, who can tax its production at a high percentage rate without seriously reducing the owner's willingness to proceed with extraction.

The sense of possession we here describe may be encountered in resource-rich local communities of all sizes and types, from the parish, through the region, right up to the populations (or their spokesmen) of subcontinental areas such as Saudi Arabia and the Persian (or Arabian) Gulf. But its manifestation in federal systems, such as the United States, or in quasi-federal systems, such as the European Community, is particularly problematical. In such systems there exist, in the shape of the states (US) or the Member States (Community), "local" governments with a comprehensive legislative and administrative apparatus and extensive regulatory and taxation powers, through which the local "sense of possession", whether felt at state level itself or in some smaller unit, may possibly find concrete expression. At the same time the integrative goals of such systems, in particular the creation of a unified market for natural resources as for other products, are peculiarly susceptible to damage from the exercise of such "local" powers. Energy-related natural resources, therefore, make a natural testing-ground for institutions aimed at promoting integration.

B. Definition of Property Rights in Energy-Resources

"Fugacious" underground resources (*i.e.* liquids and gases) pose a special challenge to a legal system. Unless *all* the geologically related reservoirs *making up an oil field* happen to belong to a single owner, ill-defined property rights in the resource will engender market failure. In the United States it was natural (if not inevitable) that the concept of ownership "*ad infernos*" would be applied to give each surface owner a property right in all oil and gas resources below his surface. It was also natural (though somewhat less inevitable) that that property right would be defined by the "Rule of Capture". The Rule entitles each owner to extract oil or gas by means of wells on his tract (and bottomed under his tract) without liability to an adjacent owner, even though particular molecules of the extracted resource may have originated in the latter's land. In effect, then, an overlying owner's property right in oil and gas does not vest until he brings it to the surface; until then his property right is defeasible by the extraction of his neighbour.

As always, such defeasible rights trigger wasteful activity. Each owner scrambles to secure his right by extracting oil or gas before his neighbour. In the ensuing struggle the owners drill too many wells, causing loss of reservoir pressure which leads to a smaller total recovery from the field than would be achieved through drilling a limited number of properly placed wells. So doing, they extract the resource prematurely, dissipating its "user" value — the value of being able to extract the oil or gas at a future time when (by virtue of greater scarcity, etc.) its value may be greater than now.

States have responded with two types of remedies, "unitisation" and specific anti-waste rules. Unitisation constitutes a reshuffling of rights so that decisions as to operation of the reservoir are made by the owners on a unified basis, taking into account all the effects that each well has on aggregate production from the reservoir. This substantially eliminates the problems giving rise to market failure. While all the owners in a reservoir may arrive at a unitisation agreement voluntarily, a state may facilitate the process by enabling owners of a specified majority of interests (say, 65 percent) to bring about unitisation despite the resistance of a substantial minority.

While unitisation aggregates the diverse interests into a unified ownership that is expected to adopt its best guess at an optimal programme of extraction, specific anti-waste rules constitute direct state constraints on development. The legislature adopting them — or authorising their adoption by an administrative agency — seeks directly to prohibit the more extreme forms of wasteful behaviour. The most important prohibitions are in the form of well-spacing rules (prohibiting, for example, more than one gas well for each 640 acre tract) and allowables (restricting the rate of extraction from each well or tract). As the persons imposing the limits are not owners of the resource, their interests do not provide as direct an incentive to decisions maximising the net value of the resource as do the interests of owners of a unitised tract. Moreover, natural concerns about the risk of arbitrary government behaviour force the controlling agency to act by more across-the-board rules than would be ideal for such distinctive resources as oil and gas reservoirs. Thus well-spacing and allowables rules seem less likely to generate optimal extraction patterns than does unitisation.

For our purposes, however, a more important distinction is that explicit state conservation rules such as well-spacing and allowables represent an opportunity by which states controlling a high proportion of a resource can improve their economic position at the expense of others. The sum of allowables clearly represents a ceiling on aggregate production in a state. Since supply constraints typically mean higher prices, the allowables system provides at least an opportunity for states to secure monopoly (or oligopoly) profits. The potential for interstate tension is clear.

C. Energy as a Vital Element in a Unified Market

Some centrifugal tendencies can easily be accommodated in the process of construction and maintenance of an economic union, by reason of the fact that they affect matters which are marginal to the operation of the union. Cultural, religious, and moral divergences may be able to subsist comfortably within even an advanced economic union, as long as these do not bear directly on important economic matters. Divergent tendencies may be tolerable even in economic matters, so long as these touch subjects of minor importance: different policies in Member States regarding the issue of commemorative postage stamps, for example, are unlikely to tear the Common Market apart. Energy, however, is a production factor of fundamental and universal importance to economic activity, which cannot be ignored or downgraded in the construction of an economic union. Energy prices enter as a significant factor into most production activities, into many service activities (transport being an outstanding example), and into private living costs. The process of integration, and the commitment to a unified market implied therein, creates a strong and widespread expectation, particularly in resource-poor areas, of equal or even privileged access to the energy resources located within the territory of the union.

Those who make and implement policy for the development of a unified market, and for the guidance of the economy at the federal or Community level, therefore, cannot treat with indifference the localising and centrifugal tendencies discussed in the two previous sections, but must confront and in some way manage them. By this we do not mean that every expression of such tendencies must be eliminated in order that an economic union may function properly. Indeed, our conclusions on federal policy in the United States tend to suggest that, even in an integrated market, a degree of local initiative may produce better results in energy management than may centralised policy-making. Our review of the limited progress of Community energy policy indicates that this is the line in policy development now being pursued even by the EC Commission. What holds good for the United States market may not necessarily apply, however, to European markets which are in a much earlier stage of integration. The substance of this essay, therefore, consists of a detailed comparison of legal integration in United States and European energy markets, on the basis of which we reach conclusions both on centralised regulation of the United States market and on the benefits and dangers of the Community's new decentralised approach. The important thing is to get the balance right: analysis of the American experience may indicate how far down the path of tolerance of centrifugal tendencies the Community can go without risking the disintegration of the elements of the common market so far put in place.

In Chapter 2 we look at state (and Member State) interventions in energy markets and at the response of federal courts and Community authorities, and examine the capacity of the United States Constitution on the one hand, and the free trade provisions of the three Community Treaties on the other, to eliminate or restrict any market-distorting or disintegrative effects such interventions may have. In Chapter 3 we turn to the formation of policy for energy resources at the federal or Community level, looking in particular at how far this operates by way of replacement, and how far by way of support, supplementation and control, of state and Member State initiatives.

Applied to the United States the approach highlights the dynamics of the relations between states and the national government, between legislatures and courts. We are struck by the ambiguity of the emerging pattern. Most particularly, vigorous federal court suppression of state interventions does not necessarily entail victory for the values in whose name those courts intervened. Sweeping aside state intrusion on a national market may help generate national action by creating a "vacuum of regulation", but the national action in turn may tend to balkanise the market. While suppression of state power almost by definition shifts power to the centre, the centre may not exercise that power to protect the unity of the national market.

Applied to the Communities, the approach discloses a quite different situation. While Member State interventions which distort trade flows within the common market are more explicitly condemned by the EC Treaties than by the US Constitution, there is in fact a high degree of toleration of such interventions in the energy sector, and little litigation, so that there is no "vacuum of regulation" to draw in Community measures. Moreover, the general style of such Community measures as have actually operated has been to support and coordinate Member State policies (including even policies that divide the market), rather than to implement comprehensive policies that substitute for Member State action. The Community energy market has not therefore been united and then balkanised like that of the United States: in large measure it remains divided along Member State lines. In a final chapter we develop and explore the significance of this contrast.

II. Background Facts and Issues

Before embarking on this programme, it may be helpful to mention some background facts and issues on which important differences exist between the United States and the European Community. Some of these matters we shall discuss later; others can be disposed of here.

A. Degrees of Self-Sufficiency in Energy-Related Natural Resources

Both the European Community and the United States are now substantial energy importers. Europe's degree of dependence on supplies from third countries is, however, much greater than the United States', as Table 1 shows. These figures relate to 1982, and it should be noted that they overstate the current dependence of both Europe and the United States on oil imports, by reason of reductions in consumption and increases in domestic production occurring since that date.

The most obvious effect of this difference in degree of dependency is that the Community could not enjoy the luxury of holding petroleum prices below world levels to the degree the United States could. To keep European prices much below world levels would have required the use of tax revenues to subsidise the imports, *i.e.*, to make up the difference between the sales price to consumers and the cost of the imports.[3] The United States, by contrast, was able

[3] Some Member States could have used economic rents from the development of state-owned energy resources for this purpose, but even for those Member States such revenues could not have financed much of a price differential for most of the period 1973-81.

Table 1
Energy Balances of EC and US, 1982 (million tons of oil equivalent)

	Crude oil	Natural gas	Solid fuels
European Community			
Production	119.84	120.76	195.46
Imports	358.89	81.37	67.13
Exports	66.49	38.67	18.96
Total energy requirement	415.20	161.50	230.66
Net imports as % of consumption	70.42	26.40	20.88
United States			
Production	486.42	422.92	552.09
Imports	211.36	23.02	0.55
Exports	3.88	1.42	69.52
Total energy requirement	690.56	438.42	470.01
Net imports as % of consumption	30.04	0.05	net exporter

SOURCE: Compiled by the authors on the basis of information contained in OECD/IEA, ENERGY BALANCES OF OECD COUNTRIES 1970/1982 (1984).

to subsidise imports heavily, relying only on revenues derived from domestic producers of crude oil through price controls and the "entitlements" programme.[4] The reader should not, however, infer that the Community's more constrained position has altogether prevented its Member States from adopting policies similarly tending to exacerbate scarcity in energy supplies.[5]

In addition, the Treaties establishing the framework of European economic integration focus more on external commercial relations than does the Constitution — understandably so in view of the Community's high level of natural resource dependency. We discuss the rather different provisions of the three Community Treaties in Chapter 3 below. As for the United States, its character as a customs union never seems to have been seriously questioned. State duties upon imports and exports are explicitly forbidden by the Constitution, and the possibility of any exception to complete and exclusive federal power over foreign trade in energy resources appears to be beyond discussion.

The difference between the overall import dependence of the United States and the Community conceals a significant feature common to the two unions.

[4] *See infra* Ch. III, at pp. 113-26.
[5] *See infra* Ch. IV, at pp. 152-53.

In both of them, as we have already hinted, domestic energy resources are very unevenly distributed. Within the integration framework, therefore, the countervailing pressures — of the resource-rich for local advantage, of the resource-poor for the equalising benefits of market integration (or rent redistribution) — are likely to be strong. In 1982, degrees of energy self-sufficiency within the European Community, as indicated by the percentage of all energy consumption represented by imports, were as shown in Table 2.

Table 2
EC Energy Self-Sufficiency — Imports as a Percentage of Consumption by Member States

Belgium	98.6	Ireland	62.3
Denmark	89.9	Italy	84.5
France	65.3	Luxembourg	93.6
Germany	51.3	Netherlands	11.2
Greece	68.0	United Kingdom	(net exporter)

SOURCE: Compiled by the authors on the basis of information contained in OECD/IEA, ENERGY BALANCES OF OECD COUNTRIES 1970/1982 (1984) and in *Table A* of the Appendix to Ch. I, *infra* p. 25.

While some caution is needed in the interpretation of these figures, they indicate the very broad range of energy resource endowments within the Community. For the United States, with 51 states, the range may be represented graphically as shown in Figure 1. Here the range is broader, though it may be noted that a relatively small group of states (13 out of 51) fall into the "exporter" category.

B. Different Structures for the Energy Supply Industry in Europe and the United States

We make no attempt to describe here in detail the structure of the energy supply industry in Europe and the United States. It would be surprising if there were not substantial differences between such structures. Spatial relationships between resources and markets differ in the United States and Europe. The United States industry has developed within a largely unified market; its European counterpart, at a distance of thirty years from the foundation of the Community, is still strongly shaped by its development within a number of separate national economic, social, administrative and legal frameworks which the process of integration has yet to break down. The simple factor of language insulates European markets in a way unknown in the United States. There is a stronger commitment to private ownership and private enterprise in the United States than in Europe. It is only this last difference in the two industrial environments that we wish to dwell on here: it bears importantly upon the

Figure 1: US Energy Self-Sufficiency — Production as a Percentage of Consumption by States

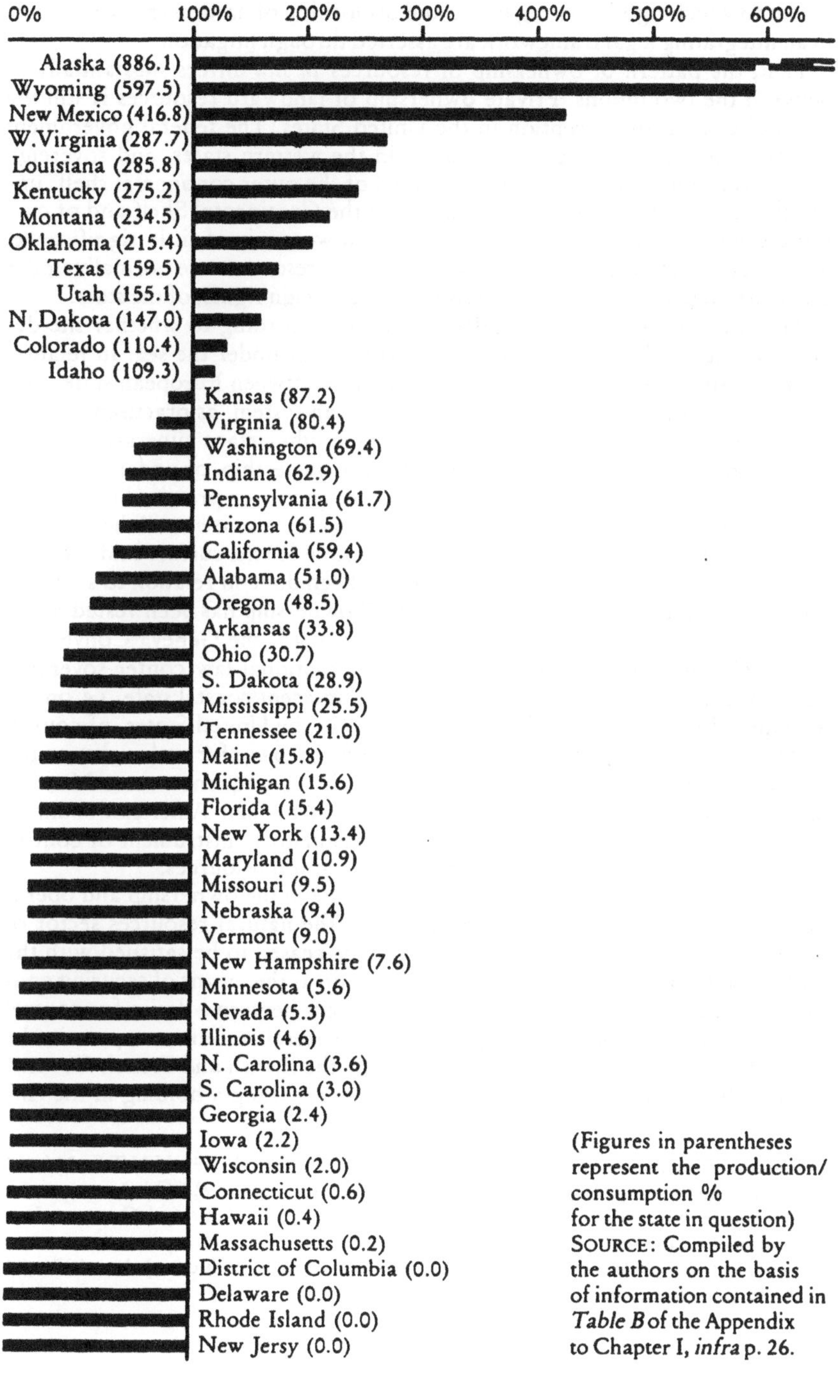

(Figures in parentheses represent the production/consumption % for the state in question)
SOURCE: Compiled by the authors on the basis of information contained in *Table B* of the Appendix to Chapter I, *infra* p. 26.

extent to which rights relating to use and disposition of natural resources within an integrating legal framework are asserted through litigation.

First, the pattern of ownership of resources *in situ* differs substantially as between the two unions. Private ownership of landward resources is still the rule rather than the exception in the United States. The federal government has extensive landholdings, particularly in the western states, and sovereign rights over mineral resources in the seabed of the outer continental shelf, under the terms of the Geneva Convention on the Continental Shelf and of customary international law.[6] State property, however, is of little significance. The reverse is the case in Europe, where energy resources are normally under the ownership, or subject to the exclusive or sovereign rights, of the state.

This position is reached by different routes according to whether the deposits are located on land in the state territory, or under the sea. In relation to landward deposits, ownership seems to differ between European states, but the result is usually the same. Under the regalian system, as practised, for example, in France and the Netherlands, rights in sub-surface minerals are separated from surface rights, and are placed in the ownership, or under the exclusive control, of the state.[7] Under the accession system, as practised in the UK, the presumption of the law is that the surface owner also owns all the minerals beneath that surface. It is, however, quite possible for the surface and sub-surface ownership to be separated, and this has been done in relation to such important minerals as coal and oil, in which all ownership has been vested in the state.[8] In relation to mineral deposits under the sea, the Geneva Convention on the Continental Shelf and customary international law confer sovereign rights of exploration and exploitation on the adjacent coastal state, *i.e.* on the individual Member States of the Community. In the United States, of course, these same provisions of international law operate to confer such rights on the federal government, not state government. In consequence most Member States of the European Community, in contrast to their American counterparts, dispose of a proprietary, as well as a regulatory, instrument of control over the use and disposition of energy-related natural resources.

Second, the degree of involvement of the state in the ownership and operation of energy industries is very different. In the natural resources sector of the economy, European states have long since ceased to be satisfied with the traditional techniques of taxation and regulation as the means of securing control and financial returns. Early signs of the now pervasive desire for control by active participation in the exploitation of natural resources appear in the search for security in imported oil supplies by the British and French Governments before and after World War I, the former taking a holding in the An-

[6] Geneva Convention on the Continental Shelf 1958, art. 2, 499 U.N.T.S. 311, 52 A.J.I.L. 834 (1958); North Sea Continental Shelf Case, [1969] I.C.J. 3.

[7] Campbell, *Principles of Mineral Ownership in the Civil Law and Common Law System*, 31 Tulane L. Rev. 303 (1957).

[8] *See* Coal Act, 1938, 1 & 2 Geo. 6, ch. 52; Petroleum (Production) Act, 1934, 24 & 25 Geo. 5, ch. 36.

Table 3
Electricity — Public Ownership

Country	Organisation	Ownership	Share of Market
UK	CEGB	100% state owned	99% conventional production 100% nuclear production monopoly of distribution
France	EDF	100% state owned	87% of production
	CNR	mixed (state/private)	6% of production
NL	11 producing companies owned by municipal or provincial authorities 94 distribution companies owned by municipal or provincial authorities		
Italy	ENEL	100% state owned	78% conventional production 74% hydro-electric production 100% nuclear production
Germany	RWE	mixed (30% owned by Länder and municipalities)	
	VEW	mixed (state/private)	45% production
	Veba	mixed (state minority share)	30% distribution

glo-Iranian Oil Company (now the British Petroleum Company), the latter participating in the formation of the Compagnie Française des Pétroles.[9] The same motive lay behind the creation in 1926, of the Italian national oil company, Agip, now a subsidiary of the Ente Nazionale Idrocarburi (ENI), and the creation in 1973-1974, via Veba AG, of a substantial government holding in the West German oil exploration company, Deminex.[10]

Of more immediate relevance here is state participation in the exploitation of domestic natural resources. This has come about in two ways: first, states have nationalised energy undertakings with the objective of securing greater control, a more appropriate framework for large-scale investment, greater production, and so on. Among nationalised enterprises of this kind we may number the Charbonnages de France, Electricité de France and Gaz de

[9] L.E. Grayson, National Oil Companies 48-75 (1981).
[10] *Id.* at 107-75.

Table 4
Nuclear Energy — Public Ownership

Country	Organisation	Ownership	Share of Market
UK	CEGB	100% state owned	100% production
	BNFL	100% state owned	manufacture of reactors monopoly of research
France	Cogema	wholly owned subsidiary of CEA	monopoly of fuel cycle process
	CEA	100% state owned	research, and control of all nuclear activities
	Framatome	34% owned by CEA	sole French manufacturer of nuclear steam supply system
NL	—	—	—
Italy	Agip Nucleare	100% owned by ENI	monopoly on acquisition of fuels
	Finmeccanica	subsidiary of IRI	sole licensee for PWRs
Germany	RWE	*see* Table 3	largest single producer

Table 5
Coal — Public Ownership

Country	Organisation	Ownership	Share of Market
UK	NCB	100% state owned	99% monopoly of production
France	CDF	100% state owned	monopoly of production
	ATIC	100% state owned	monopoly of imports
NL	—	—	—
Italy	Agip Carbone	wholly owned subsidiary of ENI	importation of coal
	ENI	100% state owned	monopoly of production
Germany	Ruhrkohle AG	mixed (state/private)	77% of production
	Saarbergwerke	100% Federal/Länder owned	16% of production
	Rheinische Braunkohle	wholly owned subsidiary of RWE (*see* Table 3)	85% lignite production

Table 6
Gas — Public Ownership

Country	Organisation	Ownership	Share of Market
UK	BGC	100% state owned	monopoly of sales until 1982
France	GDF	100% state owned	78% of sales
	SNGSO	wholly owned subsidiary of GDF	22% of sales
	Elf-Aquitaine (SNEA)	70% state owned	96% of production
NL	NAM	mixed (state minority share)	operates Groningen Concession; can take 40% or 50% participation in offshore production licences
	Gasunie	mixed (50% state share)	monopoly of sales
Italy	ENI	100% state owned	monopoly of onshore production
	SNAM	100% subsidiary of ENI	de facto monopoly of wholesale and industrial distribution
Germany	—	—	—

Table 7
Oil Production — Public Ownership

Country	Organisation	Ownership	Share of Market
UK	Oil & Pipelines Agency	100% state owned	right to acquire (in emergency) 51% of production at market price
	Britoil	mixed (state minority share)	
France	Elf-Aquitaine (SNEA)	70% state owned	
NL	—	—	—
Italy	ENI	100% state owned	exclusive production rights
Germany	—	—	—

Table 8
Oil Distribution — Public Ownership

Country	Organisation	Ownership	Share of Market
UK	—	—	—
France	Total	subsidiary of CFP	50% finished products
	Elf	subsidiary of Elf-Aquitaine	
NL	—	—	—
Italy	Agip	subsidiary of ENI	34% finished products
Germany	Aral	subsidiary of Veba (mixed with state minority interest)	25% petroleum products

France; in Britain, the National Coal Board, the British Gas Corporation (BGC), and the various electricity boards; in Italy, the Ente Nazionale per l'Energia Elettrica (ENEL); the Dutch State Mines;[11] and the state-owned Danish Natural Gas Company (DONG). Not all these enterprises, of course, are directly concerned with the production and distribution of natural resources as here defined. The role of electricity undertakings, for example, is rather one of purchase and transformation. Second, in the specific field of hydrocarbons, West European states with significant production possibilities have tended to follow the example of the Middle Eastern and North African oil states in establishing national oil companies, not to exercise a monopoly in the task of exploitation — as do the above mentioned nationalised enterprises in their respective sectors — but to provide leadership or control in an activity in which private capital (principally in the form of the major multinational oil companies) is also called upon to participate. The most significant member of this group, by reason of the richness of the oil province in which it operates, has been the British National Oil Corporation (BNOC), but mention should also be made of ENI in Italy (now the holding company of Agip), and of Elf-Aquitaine in France. Tables 3-8 (above) serve to indicate the importance of the public sector in the energy industries of the four largest Common Market countries, as well as of the Netherlands, its second significant energy producer.[12]

[11] W. Keyser & R. Windle, Public Enterprise in the European Economic Community (A Metra Oxford Study, Metra Oxford Consulting Ltd., for the EC Commission, DG IV D4, 1977). The tendency towards state participation has recently been reversed in the UK: *see* the Oil and Gas (Enterprise) Act 1982, ch. 23, the Oil and Pipelines Act 1985, ch.62, and the Gas Bill 1985, *infra* Ch. II, at n. 137.

[12] The market shares given in the tables are compiled from a variety of sources: *Germany:* W. Mönig, Schmitt, Schneider & Schurmann, Konzentration und Wettbewerb in der Energiewirtschaft (1977); L.E. Grayson, *supra* note 9; W.

In the United States, this kind of state, or even public, enterprise is rare. There is virtually no public energy enterprise outside the electricity industry. Proposals for a Federal Oil and Gas Corporation, to operate in both the upstream and downstream sectors of the industry, were laid before Congress on at least two occasions in the 1970s as a cure for alleged industry anti-competitiveness, but made little progress.[13] Even within the electricity industry, where, as we have seen, most European countries show very high levels of public participation, the share of electricity generation (including nuclear generation) accounted for by all publicly-owned utilities amounted in 1981 only to 22.2 percent, a figure which had remained almost constant over the previous ten years. This indicates that while energy utility regulation may be vigorously criticised in the United States on efficiency grounds,[14] public ownership is not perceived as an acceptable alternative means of advancing public policy goals in this area. In the United States, therefore, discussion of market integration questions can focus almost exclusively on governmental regulation of private activity. In Europe we must take account also of a powerful alternative paradigm, operation of energy industries by the state itself. We develop our argument on the significance of these contrasts in our conclusions.[15]

III. Complexity and Explicitness in Legal Frameworks

The handling of energy resource issues, as of other issues, cannot but be affected by the considerably greater complexity and explicitness displayed by the European legal framework in comparison with its American counterpart. Some of the most obvious differences in United States and European integrational practice and experience can be related to the fact that we must speak of European Communities in the plural. Though it is now popular to use the singular term "European Community" one needs always to bear in mind that competence of the Community institutions[16] — Commission, Council, Court and

KEYSER & R. WINDLE, *supra* note 11; 49 PETROLEUM ECONOMIST (Apr. issue, 1982). *Italy:* W. KEYSER & R. WINDLE, *supra* note 11, at 627-800; L.E. GRAYSON, *supra* note 9. *France:* N.J.D. LUCAS, ENERGY IN FRANCE: PLANNING POLITICS AND POLICY (1978); L.E. GRAYSON, *supra* note 9; W. KEYSER & R. WINDLE, *supra* note 11, at 169-323. *Netherlands:* W. KEYSER & R. WINDLE, *supra* note 11, at 857-99 and 900-1109. *Britain:* MONOPOLIES AND MERGERS COMMISSION, PETROL: A REPORT ON THE SUPPLY OF PETROL IN THE UNITED KINGDOM BY WHOLESALE, CMD. 7433 (1979).

13 *See* Bull, *Competition Between Public and Private Enterprises*, in LAW IN THE U.S.A. IN THE BICENTENNIAL ERA 433, 444-50 (J.N. Hazard & W.J. Wagner eds., supplement to 26 AM. J. COMP. L., 1978).

14 *See, e.g.*, G.J. STIGLER, THE CITIZEN AND THE STATE: ESSAYS ON REGULATION, ch. 5 (electricity supply)(1975).

15 *See infra* Ch. IV.

16 For general discussion of institutional competences *see* 1 INTEGRATION THROUGH LAW *passim* (1986).

Parliament — to deal with energy resource issues stems not from a single constituent document but from three separate treaties[17] which continue in existence notwithstanding the merger of the institutions of the three Communities, effected for the Court and Parliament in 1958[18] and for the Council and Commission in 1967.[19] These three Treaties, establishing respectively the European Coal and Steel Community (1951), the European Economic Community (1957), and the European Atomic Energy Community (Euratom) (1957), differ widely in scope, style and effects. We cannot, therefore, assume that the same rules govern the kinds of problems of natural resources policy with which we are concerned: different rules may apply according to whether the problem involves, for example, coal, uranium, or oil (to refer only to energy-related minerals).

The reader in the 1980s may well ask how in 1957 the Framers of the EEC and Euratom Treaties can have contemplated with equanimity the division of energy-related natural resources among three largely separate sets of institutions applying three separate treaties. Central to an answer to this question is the failure — some would say a wilful failure — on the part of the Framers to perceive the growing importance of oil imports to Western Europe, and an excessive faith in the rapid development of cheap supplies of nuclear energy.[20] What shaped the Rome Treaties was the desire to find concrete expressions of the will for European unity in vital domains. Nuclear energy was seized upon as being the most vital of all, a sector of unlimited future potential in which co-operation and integration were essential if the countries of Western Europe were to have any chance of escaping the hegemony of the United States and the threats and challenges of the Soviet bloc. The Messina Conference of Foreign Ministers of the Coal and Steel Community, which formally initiated the process of developing the Rome Treaties, concluded with a resolution one of whose explicit aims was "the creation of a common organisation to be entrusted with the responsibility and the means for ensuring the peaceful development of atomic energy..."[21] From the beginning, therefore, the integration

[17] In determining the content of Community law and the competence of Community organs it may also be necessary to look at other treaties, *e.g.*, the Treaty of Accession by which Denmark, Ireland and the United Kingdom became members of the EEC and Euratom (Brussels, 22 January 1972). These countries acceded to the ECSC Treaty under a different procedure: *compare* ECSC Treaty art. 98 *with* EEC Treaty art. 237.

[18] Convention on Certain Institutions Common to the European Communities (Rome, 25 Mar. 1957).

[19] Treaty Establishing a Single Council and a Single Commission of the European Communities (Brussels, 8 Apr. 1965).

[20] On the treatment of these questions in the development and drafting of the Rome Treaties, *see generally* N.J.D. Lucas, Energy and the European Communities 11-29 (1977).

[21] Resolution adopted by the Foreign Ministers of the States Members of the E.C.S.C. at their Meeting at Messina on June 1 and 2, 1955, in Correspondence Arising Out of the Meeting of the Foreign Ministers of the Governments of

of the nuclear energy sector was seen as a project separate from, though parallel to, the development of a general common market. It was treated separately in the *Spaak Report*,[22] and it was not surprising that it should finally have culminated in the adoption of a separate treaty.

Less easy to understand is the treatment of conventional energy sources in 1957. The Messina Conference had devoted considerable interest to conventional as well as to nuclear energy, and the Spaak Committee established an expert Commission on the subject and devoted a section of its report to it. Yet there was no amendment to the Coal and Steel Community Treaty in 1957 and energy is simply not mentioned in the European Economic Community Treaty. Neglecting the impact of continuing supplies of low-cost oil, the Committee saw only the problems of the coal-based electricity and gas supply industries, in particular their need for coordinated investment planning in the period of transition to nuclear energy as the main primary energy source. Their operation as public utilities, usually in the form of local public or private monopolies, suggested that the regime of open competition planned for the general common market might not be appropriate; and yet they were not placed under the explicit tutelage of the more dirigiste ECSC. The interest and experience of the ECSC in relation to the coal-based energy industries was recognised only by the conferment on the High Authority of the ECSC of responsibility for convening a mixed committee composed of representatives of the Member States and of the executives of the EEC and Euratom.[23]

The task of this Committee was to make studies of long-term needs and of productive investments, and to put forward proposals to the Council of Ministers for carrying them out, on an inter-governmental basis. Plainly these arrangements were not well designed to cope with the supply situation which actually ensued, in which oil and natural gas have come to account for over 63 percent of EC energy consumption, with coal and lignite reduced to 25 percent and nuclear energy, at 6 percent, just managing to outstrip geothermal and hydroelectric power as an energy source. [24] Certainly, the merger of the executives in 1967 has improved co-ordination — energy policy is now the responsibility of a single Directorate-General within the Commission — but the difficulties of working within three separate Treaty frameworks remain.

While the Community legal framework can thus be regarded as needlessly complex and fragmented, it cannot be reproached, at least in comparison with its United States counterpart, with lack of explicitness. We have stated that the

Belgium, France, the Federal Republic of Germany, Italy, Luxembourg and the Netherlands held at Messina on June 1-2, 1955, Cmd. 9525, at 7 (1955).

[22] Comité Intergouvernemental créé par la Conférence de Messine, Rapport des Chefs de Délégation aux Ministres des Affaires Etrangères (Brussels, 21 Apr. 1956) (the *Spaak Report*).

[23] *See* L. Lister, Europe's Coal and Steel Community: An Experiment in Economic Union 335-36 (1960); 1957 J.O. (CECA) 574 (7 Dec. 1957).

[24] Figures calculated from OECD/IEA, Energy Balances of OECD Countries 1970/1982 (1984).

EEC Treaty makes no reference to energy; nor, of course, does the US Constitution. An important part of our discussion, however, on control of local initiatives in relation to energy-resources, raises issues in the application of Treaty and Constitutional provisions regarding the free movement of goods. On this general theme the Community Treaties are much more clear and explicit than the US Constitution. As between the Member States, they prohibit import and export duties and duties equivalent in effect thereto, quantitative restrictions and measures of equivalent effect, and certain other measures distorting the pattern of inter-state trade such as discriminatory internal taxation, state aids[25] and manipulation of the behaviour of state enterprises. With the exception of state duties on imports and exports,[26] no such specific prohibitions are to be found in the US Constitution: the principle of free movement of goods needs to be inferred from the Constitution as a whole and in particular from the conferment on Congress of the power "to regulate Commerce ... among the several states." The Supreme Court has held that such power, while not automatically precluding any state regulation that affects interstate commerce, does preclude some such regulation even in the absence of conflicting Congressional action. As the text of the Constitution provides no express criteria for distinguishing between permissible and impermissible state regulations, the Court has had to develop its own tests. This can lead to considerable uncertainty, as we shall see, when unfamiliar situations, such as state subsidies or the activities of state enterprises, are brought to the attention of the courts.[27] In regard to explicitness of provision, therefore, the position in relation to freedom of trade is the reverse of that observed in relation to supremacy of federal law, clearly proclaimed by the US Constitution but in the Communities based only on judicial deduction and even now seriously contested.[28]

IV. Geographical Aspects of Federal and Community Jurisdiction

Finally, this introduction is a convenient place to signal the absence, in the United States, of any controversy corresponding to that of whether the EEC Treaty extends to offshore natural resources. It is clear, in United States law,

[25] EEC Treaty arts. 92-93; ECSC Treaty art. 4(c). Note the treatment, by the US Supreme Court, of a state aid as an element of market forces, in Hughes v. Alexandria Scrap Corp., 426 U.S. 794 (1976) (discussed *infra* Ch. II).

[26] US CONST. art. 1, § 10. *See* Michelin Tire Corp. v. Wages, 423 U.S. 276 (1976). Interstate transfers do not constitute imports or exports under this section: Woodruff v. Parham, 75 U.S. 123 (1869).

[27] *See infra* Ch. II *passim.*

[28] *See generally* Cappelletti & Golay, *Judicial Review, Transnational and Federal: Its Impact on Integration*, in I/2 INTEGRATION THROUGH LAW 261, 262-64, 279, 309-15 (1986).

that jurisdiction and control over the resources of the outer continental shelf rest in the federal government. In most important respects, therefore, the resources of the outer continental shelf may be assimilated to resources found in federal lands, a kind of property which has no counterpart in the European Community structure. In the Communities, as we have seen, there is no question of common property in offshore resources, which clearly are attributed to the respective coastal states. The question is rather one of whether, in exercising these sovereign offshore rights, Member States are bound by the integrative requirements of the Treaties. If they are not, then the integrative effects of the Community structures are significantly limited in relation to natural resources. In terms of our theme, oil is the important resource: no large deposits of coal or uranium have yet been found within the offshore jurisdiction of Member States.[29] We have already noted that under international law coastal states enjoy sovereign rights to the resources of the adjacent continental shelf. Member States with claims to areas of the north-west European shelf by virtue of their position as coastal states are Denmark, West Germany, Belgium, France, Ireland and, preeminent by reason of the importance of the resources involved, the Netherlands and the United Kingdom. A comparable position is enjoyed by France, Italy and Greece in relation to parts of the Mediterranean beyond their territorial seas. The task of delimiting the continental shelf as a whole and the areas of it appertaining to adjacent or opposite coastal states has everywhere been complex, and even as between neighbouring partners in the European Communities has not always been accomplished by agreement. Settlement of boundaries as between Denmark, West Germany and the Netherlands required a reference to the International Court of Justice,[30] the United Kingdom-France borderline in the Channel and Western Approaches had to be determined by arbitration,[31] and there is continuing disagreement between the United Kingdom and Ireland about delimitation in the Western Approaches and in the North East Atlantic which will need to be settled by the same means.

The specifically European problem raised by continental shelf resources, however, is different: it is whether such resources fall within the contemplation of the Treaty at all. Given that rights over continental shelf resources are enjoyed by states which are subject to the Treaty, it might, on first impression, be assumed that the Treaty would apply. This assumption has, however, been contested by some Member States, and in the twenty-six years of the Treaty's life, no clear decision on this aspect of its geographical scope has been reached

[29] Note that there may be differences in the geographical scope of the Treaties. The ECSC Treaty applies only to the European territories of the Member States (art. 79); the Euratom Treaty applies, save as otherwise provided, only to such territories and to non-European territories under their jurisdiction (art. 198). For the EEC Treaty, *see infra* text accompanying notes 32-50.

[30] North Sea Continental Shelf Case, [1969] I.C.J. 3.

[31] Anglo-French Continental Shelf Case (1977-78), CMD. 7438 (1979), 18 I.L.M. 397 (1979).

either by agreement[32] or by judicial decision.[33] While the Commission has put the case for a broad interpretation of the Treaty in this regard in a Memorandum published in 1970,[34] the arguments adduced by some Member States to the contrary have not been published. We may surmise, however, that they hinge on the proposition that since the continental shelf does not form part of the territory of the Member States and in fact lies outside their customs jurisdiction,[35] it is, therefore, not covered by article 227 of the EEC Treaty, which simply provides that the Treaty shall apply to "the Kingdom of Belgium, the Kingdom of Denmark, etc." The effect of this argument, it should be noted, is not to isolate continental shelf resources entirely from the regime of the Treaty, as is the case with defence materials covered by article 223. Once the resources are brought to land within the coastal state (or, for that matter, elsewhere in the Community) they are fully subject to the Community regime on free movement of goods. The question is only whether the Treaty can govern their control and disposition while *in situ* and in the process of being brought to land. If it cannot, then the problems which have arisen in the past and may continue to arise in the future regarding British policy for the production and disposition of North Sea oil cease to be troublesome, at least for the lawyer: the Community simply has no legal competence in the matter. In the sections that follow, however, it is assumed that there is Community competence. This is partly because, while reserving their position on the geographical question, Member States with continental shelf resources — in particular the United Kingdom — have clearly designed certain elements of their legal regime on

[32] For an example of studied ambiguity on this point, *see* EEC Council Directive 69/82 of 13 March 1969, J.O. L 68, 19 Mar. 1969, p. 4, 1969 O.J. (spec. Eng. ed.) 111, on freedom of establishment in regard to exploration for petroleum and natural gas. Despite attempts by the European Parliament to secure the insertion of a clear formula (*see* Wenger, *La CEE et le plateau continental*, 1971 R.M.C. 184), the Directive simply states that it applies to the same geographical area as that covered by EEC Council Directive 64/428 of 7 July 1964, 1964 J.O. 1871 (23 July 1964), 1963-1964 O.J. (spec. Eng. ed.) 151, on freedom of establishment in mining and quarrying. That Directive, however, does not specify the area to which it applies.

[33] A reference which would have led to a decision of the Court of Justice on this point was made by the British National Insurance Commissioner in *Re* the "Key Gibraltar" Oil Drilling Rig, [1979] 1 C.M.L.R. 362, but the case was settled at a late stage by the Department of Health and Social Security and the reference withdrawn.

[34] Doc. SEC(70) 3095 final (18 Sept. 1970), *reprinted in* 10 I.L.M. 202 (1971). For comments *see* Vignes, *The EEC and the Law of the Sea*, in 3 New Directions in the Law of the Sea, Collected Papers 335-47 (R. Churchill, K.R. Simmonds & J. Welch eds. 1973); Wenger, *supra* note 32, at 189; A. Wenger, Pétrole et gaz naturel en mer du Nord 54-62 (1971).

[35] Products of the continental shelf may, however, be *deemed* to have originated in national territories: *see* the legislation cited in Wenger, *supra* note 32, at 185; and for the UK *see* Customs and Excise Duties (General Reliefs) Act 1979, ch. 3, § 14(2)-(4), discussed *infra* in Ch. II, at notes 61-62 and accompanying text.

the assumption that such competence might be found to exist;[36] and partly because developments in relation to another extra-territorial natural resource, offshore and deep water fisheries, would seem to have weakened the case against the possibly extra-territorial application of Treaty rules almost to the point of collapse.

In relation to fisheries the relevant international law concept is not the continental shelf, but the exclusive fishery zone, that is to say, an area of water, rather than of the seabed beneath it. While at one time coastal states did not normally claim exclusive rights over sea fisheries beyond the outer limit of their territorial sea, increasing competition for fishery resources and associated problems of excessive exploitation have led coastal states to extend their claimed exclusive fishing limits, both in concert, as in the London Fisheries Convention of 1964,[37] and unilaterally. The new Convention on the Law of the Sea contemplates the possession by each coastal state of an exclusive economic zone extending for 200 miles from the baseline of the territorial sea.[38]

Article 38(1) of the EEC Treaty defines "agricultural products" so as to include the products of fisheries. Fish are also included in the list of products referred to in article 38(3) as being subject to the provisions of articles 39-46, on the common agricultural policy. This means that the Member States were under an obligation, by virtue of articles 40(3) and 43, to establish a common organisation of the fisheries market. Among the objectives specified for such common market organisations are the increase of agricultural productivity by ensuring the rational development of production,[39] and the avoidance of any discrimination between producers and consumers within the Community.[40] The Member States, then six, agreed to establish such a common organisation for fisheries, under the name of a common fisheries policy, and first laid down its broad lines by regulations adopted in 1970 whose principal provisions were re-enacted in identical terms by Council Regulation 101/76.[41] Regulation 2141/70[42] laid down a common structural policy for the fishing industry which has been completed by the common fisheries policy Regulations adopted in January 1983.[43] The vital provision of Regulation 2141/70, from the point of view of the extent of application of Community law, is article 2,

[36] *See, e.g., infra* Ch. II, pp. 59-64.
[37] For the text *see* 1 New Directions in the Law of the Sea, Documents 41 (S.H. Lay, R. Churchill & M. Nordquist eds. 1973).
[38] United Nations Convention on the Law of the Sea, *opened for signature* 10 Dec. 1982, 21 I.L.M. 1261 (1982).
[39] EEC Treaty art. 39(1)(a).
[40] *Id.* art. 40(3).
[41] O.J. L 20, 28 Jan. 1976, p. 19.
[42] J.O. L 236, 27 Oct. 1970, p. 1, 1970 O.J. (spec. Eng. ed.) 703.
[43] Council Regulations (EEC) 170-181/1983, O.J. L 24, 27 Jan. 1983, p. 19; and 198/1983, O.J. L 25, 27 Jan. 1983, p. 32.

which requires Member States to ensure equality of access[44] to the maritime waters coming under their sovereignty or within their jurisdiction for fishing vessels from all Member States. The maritime waters referred to are "those which are so described by the laws in force in each Member State" (article 2(3)). Subject to this requirement the Regulation envisages the maintenance in force for the time being of Member States' laws and administrative rules and regulations governing fishing in their maritime waters, but it also contemplates (and article 102 of the 1972 Act of Accession created an obligation in this respect[45]) the adoption by the Council of conservation measures where there is a risk of over-fishing in the maritime areas referred to.[46]

It appears clear from the wording of article 2 that this Regulation was intended to have effect outside the territories of the Member States, that is to say, not only in their territorial waters, but also in any exclusive fishing zones claimed by them, under their laws in force, outside such waters. In 1970 these zones in no case extended beyond 12 miles from the baseline of any Member State's territorial sea: but with effect from 1 January 1977 the Member States, "acting in concert", extended their exclusive fishery zones to 200 miles from such baselines.[47] That the reflection of this decision in national laws must lead to the extension of the maritime waters subject to the Community fisheries policy was contested by Ireland before the European Court of Justice in 1978.[48] Ireland's case was based on a narrow interpretation of article 2(3) of the Regulation; it did not argue that to extend the fisheries regime beyond territorial waters was to exceed the geographical scope of the Treaty. In rejecting Ireland's contentions, however, the Court of Justice seemed to go out of its way to use language indicative of the applicability of the Treaty to any area over which Member States claim sovereignty or jurisdiction:[49] to the continental shelf, in other words, as well as to fisheries zones. The fishery regulations, it stated, apply in principle to the same geographical area as the Treaty itself, and must be understood as referring to the limits of the field of application of Community law in its entirety, as that field may at any given time be constituted. "Consequently," it continued, "the reference [in article 2(3)] to the 'laws in force' in the various Member States as describing the maritime waters coming under their sovereignty or within their jurisdiction must be interpreted as

[44] For an examination of a disguised restriction on equality of access, which also expresses the attitude of the Court of Justice to measures discriminatory in fact though not on their face, *see* Case 61/77, Commission v. Ireland, [1978] E.C.R. 417, 451-53.

[45] This obligation was to have been fulfilled by 31 Dec. 1978 (*see* Joined Cases 185-204/78, Van Dam en Zonen, [1979] E.C.R. 2345), but Member States were unable to agree on the details of a common fisheries policy until 1983: *see* BULL. EC 1-1983, pts. 1.1.1 to 1.1.10.

[46] Regulation 2141/70, art. 5.

[47] *See* BULL. EC 10-1976, pts. 1501 to 1505.

[48] Case 61/77, Commission v. Ireland, [1978] E.C.R. 417.

[49] *Id.* at 445-46.

referring to the laws applicable from time to time during the period of validity of the regulation concerned."[50] In the light of this it would now seem all but impossible to argue successfully for the strict territorial restriction of the application of the EEC Treaty, whether in the fisheries field or, by reason of the generality of the Court's remarks, in relation to other natural resources.

[50] *Id.* at 446, para. 48.

Appendix to Chapter I

In the compilation of the data presented in Tables A and B and summarised in Table 2 (at p. 8) and Figure 1 (at p. 9) two different measurements of energy production are used. The EC Table is based on the latest statistics provided by the OECD/IEA and this source uses the MTOE measurement. The Table includes production figures relating to nuclear electricity and excludes production figures relating to uranium.

Both the US Table and Figure 1 are based on American sources which employ the BTU measurement; these sources exclude production figures based on nuclear electricity and include in their place estimated uranium production data.

Table A
Production and Consumption Data for the EC Member States, 1982

Country	Resources in MTOE							
	Natural Gas	Coal	Crude Oil	Hydro*	Nuclear	Total Production	Total Consumption	Production as % of Consumption
Belgium	0.03	4.69	—	0.23	3.50	8.45	41.70	20.25
Denmark	—	0.41	1.69	0.01	—	2.11	17.59	11.99
France	5.53	13.98	2.48	16.06	24.33	62.38	184.92	33.73
Germany	12.98	94.37	4.99	4.39	14.20	130.93	254.75	51.40
Greece	—	3.67	1.04	0.80	—	5.51	15.91	34.63
Ireland	1.69	1.31	—	0.27	—	3.27	8.63	37.90
Italy	12.70	1.46	1.79	10.46	1.52	27.93	134.75	20.72
Luxembourg	0.28	0.02	1.08	0.11	—	1.49	3.11	47.90
NL	55.49	0.29	1.95	—	0.87	58.60	56.00	104.64
UK	32.37	75.27	105.89	1.26	9.82	224.61	196.71	114.18

* Hydro figures include both geothermal and solar energy
SOURCE: Compiled by the authors on the basis of information contained in OECD/IEA, ENERGY BALANCES OF OECD COUNTRIES 1970/1982 (1984).

Table B
Production and Consumption Data for the US, 1982[a]

	Natural gas Marketed Production[b]		Petroleum production of crude oil[c] (including lease condesate)		Coal[d]		Uranium[e]		Hydro-electric[f]	Total production[g]	Total con-sumption[h]	Production as % of consump-tion
	MCF	Trillion BTU	1,000 Barrels	Trillion BTU	1,000 Short tons	Trillion BTU	1,000 tons	Trillion BTU	Trillion BTU	Trillion BTU	Trillion BTU	%
Alabama	75,003	77.1	20,041	0.1	26,556	638.4	—	—	112.4	828.0	1,623.9	51.0
Alaska	264,364	271.8	618,910	3,351.4	883	15.4	—	—	5.9	3,644.4	411.3	886.1
Arizona	99	0.1	335	1.8	12,364	260.8	0,8[i]	(181.7)	73.4	336.1	841.5	39.9
										(517.8)		(61.5)
Arkansas	124,611	128.1	18,849	102.1	161	2.7	—	—	22.1	255.0	754.3	33.8
California	383,977	394.7	401,572	2,174.5	—	—	—	—	525.9	3,095.1	5,208.7	59.4
Colorado	209,892	215.8	30,545	165.4	18,318	359.6	0.8[i]	(181.7)	17.3	758.0	851.4	89.0
										(939.7)		(110.4)
Connecticut	—	—	—	—	—	—	—	—	3.9	3.9	684.8	0.6
Delaware	—	—	—	—	—	—	—	—	—	—	197.4	0.0
Dist. of Col.	—	—	—	—	—	—	—	—	—	—	80.1	0.0
Florida	22,515	23.1	25,626	138.8	—	—	0.8[i]	(181.7)	2.7	164.6	2,245.5	7.3
										(346.3)		(15.4)
Georgia	—	—	—	—	14	0.3	—	—	38.2	38.5	1,616.6	2.4
Hawaii	—	—	—	—	—	—	—	—	0.9	0.9	217.6	0.4
Idaho	—	—	—	—	—	—	0.8[i]	(181.7)	121.4	121.4	277.3	43.8
										(303.1)		(109.3)
Illinois	1,162	1.2	27,710	150.0	—	—	—	—	1.3	152.5	3,286.2	4.6
Indiana	233	0.2	6,213	33.6	60,275	1,312.8	—	—	4.5	1,351.2	2,147.3	62.9
Iowa	—	—	—	—	566	10.3	—	—	9.6	19.9	895.1	2.2
Kansas	429,597	441.6	70,525	381.9	1,412	25.1	—	—	0.1	848.7	973.3	87.2
Kentucky	51,924	53.4	7,349	39.8	150,215	3,424.9	—	—	35.0	3,553.1	1,291.2	275.2
Louisiana	6,171,627	6,344.4	455,326	2,465.6	—	—	—	—	—	8,810.0	3,082.3	285.8
Maine	—	—	—	—	—	—	—	—	61.7	61.7	390.6	15.8

Maryland	36	*	—	—	3,817	94.7	—	—	14.0	108.8	1,001.6	10.9
Mass.	—	—	—	—	—	—	—	—	2.6	2.6	1,160.1	0.2
Michigan	167,231	171.9	31,462	170.4	—	—	—	—	35.1	377.4	2,422.4	15.6
Minnesota	—	—	—	—	—	—	—	—	65.2	65.2	1,163.6	5.6
Mississippi	—	—	33,047	178.9	—	—	—	—	—	179.0	702.2	25.5
Missouri	—	—	202	1.1	5,341	114.3	—	—	17.3	132.7	1,400.6	9.5
Montana	56,517	58.1	30,921	167.4	27,890	474.4	—	—	114.3	814.2	347.2	234.5
Nebraska	2,280	2.3	6,872	37.2	—	—	—	—	12.7	52.3	550.3	9.4
Nevada	—	—	613	3.3	—	—	—	—	14.9	18.2	341.3	5.3
New Hampshire	—	—	—	—	—	—	—	—	13.1	13.1	171.3	7.6
New Jersey	—	—	—	—	—	—	—	—	–2.3[i]	–2.3[i]	1,845.4	–0.1
New Mexiko	991,178	1,018.9	71,024	384.6	19,944	364.6	3.8	872.2	0.8	2,641.1	633.7	416.8
New York	15,887	16.3	834	4.5	—	—	—	—	426.0	446.8	3,485.1	13.4
North Carolina	—	—	—	—	—	—	—	—	56.6	56.6	1,569.5	3.6
North Dakota	53,818	55.3	47,271	256.0	17,855	237.3	—	—	27.0	575.6	391.5	147.0
Ohio	138,391	142.3	14,571	78.9	36,490	843.3	—	—	0.1	1,064.6	3,465.3	30.7
Oklahoma	1,934,412	1,988.6	158,621	858.9	4,797	81.8	—	—	21.9	2,951.2	1,370.4	215.4
Oregon	3	*	—	—	—	—	—	—	473.5	473.5	975.7	48.5
Pennsylvania	121,111	124.5	4,282	23.2	79,359[k]	1,931.2	—	—	19.1	2,098.0	3,397.9	61.7
					(4,588)[m]							
Rhode Island	—	—	—	—	—	—	—	—	—	—	126.4	0.0
South Carolina	—	—	—	—	—	—	—	—	25.4	25.4	858.3	3.0
South Dakota	2,331	2.4	1,237	6.7	—	—	—	—	56.8	65.9	227,9	28.9
Tennessee	2,976	3.1	1,132	6.1	7,450	176.0	—	—	102.3	287.5	1,371.4	21.0
Texas	6,486,817	6,649.9	925,296	5,010.5	38,818	581.5	2.2	529.5	10.7	12,782.2	8,013.4	159.5
Utah	94,255	96.9	22,240	120.4	17,029	393.0	0.8[i]	(181.7)	10.7	513.8	461.2	111.4
										(715.5)		(155.1)
Vermont	—	—	—	—	—	—	—	—	10.9	10.9	120.6	9.0
Virginia	6,880	7.1	49	0.3	39,778	993.3	—	—	9.8	1,010.4	1,256.3	80.4
Washington	—	—	—	—	4,164	67.5	0.8[i]	(181.7)	971.5	1,039.0	1,759.4	59.1
										(1,220.7)		(69.4)
W. Virginia	150,850	155.1	3,227	17.5	128,540	3,142.8	—	—	11.7	3,327.1	1,156.4	287.7
Wisconsin	—	—	—	—	—	—	—	—	25.4	25.4	1,244.7	2.0
Wyoming	424,657	436.5	118,300	640.6	108,361	1,879.0	2.7	623.0	8.9	3,588.0	600.5	597.5

Notes to Table B

[a] Conversion factors are from STATE ENERGY DATA REPORT, CONSUMPTION ESTIMATES, 1960-82, DOE/EIA-0214 (84).

[b] NATURAL GAS ANNUAL, 1982, DOE/EIA-0131 (82).

[c] PETROLEUM SUPPLY ANNUAL, 1982, DOE/EIA-0340 (82/2).

[d] QUARTERLY COAL REPORT, Oct./Dec. 1983, DOE/EIA-0121 (84/4Q). Conversion factors for coal are assumed to equal the State Energy Data System conversion factors for bituminous coal and lignite consumed by electric utilities for each state. Production of anthracite in Pennsylvania is converted to BTUs using an anthracite conversion factor.

[e] STATISTICAL DATA OF THE URANIUM INDUSTRY, Jan. 1, 1983, GJO-100 (83). BTUs produced by each state from uranium is calculated by multiplying the total BTUs consumed in the US from nuclear power by each state's percentage of total US uranium production.

[f] It is assumed that consumption equals production in each state. A negative number results from pumped storage for which, overall, more electricity is expended than is created, to provide electricity during peak demand periods.

[g] For Arizona, Colorado, Florida, Idaho, Utah and Washington, top figure is total production not including the production of uranium. Figure in parenthesis includes 181.7 trillion BTUs, assumed to be the production of uranium for each state (*see infra* note i).

[h] Does not include geothermal energy or wood and waste.

[i] Individual data withheld for Arizona, Colorado, Florida, Idaho, Utah and Washington to avoid disclosure of individual company data. It is assumed that each state produced an equal proportion of the 35% of US uranium production attributed to all six states, resulting in 800 tons or 181.7 trillion BTUs per state.

[j] This net negative production figure reflects the fact that New Jersey possesses pumped storage facilities which provide electricity during peak periods; in this process more electricity is expended than is created. The consequent negative percentage figure is recorded as zero on the bar chart (Figure 1, *supra* at p. 9).

[k] Total production in Pennsylvania.

[m] Production of anthracite.

Chapter Two

State Interventions in Energy Markets

I. Introduction

State interventions are here grouped in terms of primary purposes: wealth redistribution, on the one hand, and efficiency through correction of market failures on the other. This grouping is only an organisational device which permits us to compare the different structures and problems encountered in the United States and Europe. It should not be taken as a claim that the purpose under which a specific action is classified was its exclusive, or even necessarily its major, goal. Our classification of a state intervention under a particular purpose simply expresses our belief that the measure would not have passed (in substantially its final form) but for its expected tendency to advance that purpose.[1] Specifically, it does not of itself imply any judgment as to the constitutionality or legality of the action in question.[2]

In fact the constitutional and Treaty tests for the legitimacy of state intervention are not couched in terms of efficiency and wealth distribution. In the United States, the principal constitutional instrument of unification of the market has been article I, section 8, clause 3, which authorises Congress to regulate "commerce ... among the several States." The Supreme Court has found in this authorisation a "negative commerce clause", a concept it uses to strike down state legislation perceived as imposing an undue burden on interstate commerce, even in the absence of any congressional action.[3]

[1] In some instances it means even less: *see, e.g.*, the discussion of Nebraska's limits on water exportation, *infra* text accompanying notes 282-84.

[2] The problem of judicial treatment of legislative purpose is discussed below. Reliance on inferences of legislative purpose for determining the constitutionality of state action risks injury to values central to American political life: the federal idea (that states should have substantial powers over their fates); and the democratic idea (that legislative bodies should be responsible to the electorate, at least formally). This study's attributions of purpose do not mean that we think that courts could legitimately make the same attribution as a basis for invalidating the statute.

[3] *See generally* Jacobs & Karst, *The "Federal" Legal Order: The U.S.A. and Europe Compared — A Juridical Perspective*, in I/1 Integration Through Law 169, 230-31 (1986); Kommers & Waelbroeck, *Legal Integration and the Free Movement of Goods: The American and European Experience*, in I/3 Integration Through Law 165, 168-97 (1986).

In the Community the commitment to free trade is both more explicit and more complex. The creation of a common market wherein goods, services, labour and capital will move without restriction or discrimination between the territories of the Member States is an element of the economic construction essayed by each of the Treaties. In the ECSC and the EEC it is a primary element, mentioned, indeed, as the first of the tasks of the Community.[4] In Euratom, by contrast, the idea of a common market appears as a subsidiary one. Member States agreed to create a nuclear common market, containing most of the same basic freedoms as in the EEC and ECSC markets, in particular provisions for the free movement of goods, including the raw materials of nuclear energy.[5] In relation to such materials, however, the mere removal of barriers to the free movement of goods was not seen as sufficient to ensure one objective of the Community, that is "that all users in the Community receive a regular and equitable supply of ores and nuclear fuels",[6] an objective which may be set alongside that of the ECSC to "ensure that all comparably placed consumers in the common market have equal access to the sources of production".[7] The particular conditions of the nuclear industry led the drafters of the Euratom Treaty to the view that this objective could not be properly reached through a free market solution alone. One such condition was the special security requirements associated with the holding, processing and transfer of nuclear fuels; another, the expectation that supplies of raw materials would continue to be difficult to obtain.[8] In consequence the Framers opted for the device of a central supply agency (the Euratom Supply Agency, ESA) with monopoly rights over the purchase and distribution of what the Treaty describes as ores, source materials and special fissile materials.[9] The details of this uniquely centralised solution merit separate treatment, notwithstanding the fact that for various reasons, which we shall examine, the Agency has never exercised its Treaty powers to the full.[10]

In confronting state interventions with these constitutional and Treaty provisions, we lay particular stress on state entrepreneurial involvement in resource development, and on state ownership of natural resources *in situ*. Though these factors have played little role in energy cases litigated in the United States, they are of major importance in the Community context. The significance of this issue emerges piecemeal, so it may help to note in advance the central propositions of the two systems regarding state ownership and state enterprise. In the United States, the central proposition is that a state may *con-*

[4] EEC Treaty art. 2; ECSC Treaty art. 2.

[5] Euratom Treaty arts. 1, 2 & 92-100.

[6] Euratom Treaty art. 2(d).

[7] ECSC Treaty art. 3(b).

[8] *See* Euratom Treaty arts. 28 & 29. The latter expectation did not turn out to be correct: *see* J.G. POLACH, EURATOM, ITS BACKGROUND, ISSUES, ECONOMIC IMPLICATIONS (1964).

[9] Euratom Treaty arts. 53-63.

[10] *See infra* Ch. III, at pp. 126-31.

duct an enterprise in accordance with standards that it could not lawfully, as a regulatory authority, *impose* upon private firms. Indeed, the Supreme Court has formulated the doctrine as an absolute, purporting to put such conduct above judicial scrutiny under the commerce clause. This proposition is subject to two variations. The first variation is that a state may not enjoy this immunity, or may enjoy it to a far lesser degree, in its management of natural resources owned by it *in situ*.[11] The second is that a state probably does not enjoy the immunity at all where its "ownership" of natural resources is a fictional shorthand for the resource's *not* being subject to a conventional system of fully specified property rights, *i.e.*, where the resource is, like the air, "common to all and the property of none".[12] In the Community, by contrast, a Member State may not use its own enterprises to achieve results which could not be legitimately achieved by regulations and state enterprises are subject to general Treaty rules; but limited exceptions exist for state monopolies and services of general economic interest.[13]

II. Redistributive Interventions

Location of significant energy-resources in a state provides that state with redistributive opportunities of several types. First, the state may seek to enable citizens owning and producing the resource to enjoy a measure of monopoly profits at the expense of consumers, preferably abroad or out-of-state, by facilitating some form of cartelisation. Obviously the success of such a venture turns critically on the ability of consumers to find substitutes. Thus, if State X created a cartel of its oil producers, the cartel's ability to secure monopoly profits would depend upon either (1) consumers' having few substitutes for State X's oil or (2) State X's bringing enough other states into the cartel to limit consumers' chances of obtaining substitutes.

Second, a state might try to transfer wealth from consumers (again preferably abroad or out-of-state) to the state treasury. A tax on production of the resource might achieve this, but its ability to do so would depend upon the same requirements as for successful cartelisation described above: consumers' having little access to substitutes.

Third, a state might seek to transfer wealth from owners and producers of the resource to others. Here its ability to do so would turn in large part on the

[11] For discussion of the basic doctrine and its first variation, *see infra* at pp. 45-56.

[12] United States v. Gerlach Live Stock Co., 339 U.S. 725, 744-45 (1950). *See generally* Note, *Federal Nonreserved Water Rights*, 48 U. Chi. L. Rev. 758, 770-72 (1981). For discussion of this second variation, *see infra* text accompanying notes 271-84.

[13] EEC Treaty arts. 37 & 90 (discussed *infra* notes 129-36 and accompanying text).

availability of economic rent — returns to owner-producers[14] in excess of those necessary to elicit production. But under market conditions the price will equal the sum of (1) marginal extraction cost plus (2) marginal "user cost", *i.e.*, the cost of foregoing the return from extraction at the next most lucrative time for extraction.[15] Economic rents, essentially user cost, will thus accrue to owner-producers of both the marginal and the inframarginal units.[16] Such rents present an attractive opportunity for redistributive activity, a chance to collect the golden eggs without killing the goose. (Minerals presumably will continue to be extracted if the price is high enough to cover extraction costs, including a normal profit.) If the state knew every producer's cost schedule it could tax the entire rent without reducing production.[17] Alternatively, a state might seek to shift the economic rent from owners and producers of the resource to in-state consumers, by granting them preferential access to production.[18]

Each of the various state enactments under review in this section appears to involve a redistribution of one of these types.

[14] If ownership and production of the resource were entirely separated and markets reasonably competitive, presumably owners would be able to enjoy the economic rents (if we put aside the effect of state interventions). For various reasons, including risk distribution, owners and producers in fact enter into arrangements under which the producer becomes effective owner of a portion of the resource. In this discussion of the distribution of economic rent we do not distinguish the two interests.

[15] *See, e.g.*, S.L. McDonald, Petroleum Conservation in the United States 76-84 (1971). This assumes a regime of adequately specified property rights, *see infra* at pp. 85-94.

[16] Howe divides these rents into two types: (1) "uniqueness", or Ricardian rents, which accrue only to those reserves that are superior (less costly to extract) to others being extracted at any time but which are nonetheless being extracted at that time; and (2) "scarcity rents", representing the present value of foregoing future use of even the marginal reserve. *See* C.W. Howe, Natural Resource Economics: Issues, Analysis and Policy 75-79 (1979). Under competitive conditions, the market price of the marginal reserve *in situ* would be its scarcity rent; the market price *in situ* of any superior reserve would comprise both its scarcity rent and its uniqueness rent.

Rents of this sort, unlike monopoly rents, by no means imply inefficient use of the resource in question. *See* S. Breyer, Regulation and Its Reform 21-23 (1982).

[17] As changes in the economic rent constitute the mechanism by which the market allocates a resource over time (*see, e.g.*, Williams, *Running Out: The Problem of Exhaustible Resources*, 7 J. Legal Stud. 165, 167-81 (1978)), any such "perfect" capture of economic rent would destroy the market as an intertemporal allocator.

[18] Alaska's effort to require persons extracting oil owned by the state, and persons contracting with them, to discriminate in hiring in favour of Alaska residents, (discussed *infra* text accompanying notes 124-27), is akin to these forms of redistribution: there the state tried to redistribute economic rents accruing to itself as owner to residents who would benefit from the hiring preference.

A. Direct Export Restrictions

1. In the United States

Early in the twentieth century two natural gas producing states sought to exploit that advantage by restricting exports. Oklahoma enacted a statute prohibiting formation (or operation in Oklahoma) of a corporation that included among its purposes the construction of interstate gas pipelines, thus effectively barring such construction.[19] A clear effect of the statute (if not its purpose) would have been to free Oklahoma natural gas consumers from the price-raising effect of purchases by out-of-state users. This would transfer some of the benefits of Oklahoma's favourable energy situation from the owners of those energy resources to Oklahomans who used it either for production or for their own consumption. West Virginia followed suit with a statute creating a regulatory commission and instructing it to give preferential treatment to in-state consumers over out-of-state ones. In *Oklahoma v. Kansas Natural Gas Co.*[20] and *Pennsylvania v. West Virginia*,[21] the Supreme Court found the statutes in violation of the "negative commerce clause" — the doctrine that the Constitution's affirmative grant to Congress of authority over interstate commerce has the collateral effect of precluding state laws which excessively intrude upon the constitutional commitment to a unified national market.[22]

Both states sought to justify the restrictions in terms of the purpose of "conserving" limited and exhaustible local supplies of natural gas. But the Court found the discriminatory character of the state's purported conservation efforts to be fatal. In *Oklahoma*, for example, the Court said that the statute "selects its [the natural gas's] market to reserve it for future purchasers and use *within the State*" (emphasis added).[23] If Oklahoma's theory were accepted, then states with coal and timber could play the same game: "Both of those products may be limited in amount, and the same considerations of public welfare which would confine gas to the use of the inhabitants of a State would confine them to the inhabitants of the State."[24]

The states might well have argued (indeed, they may implicitly have done so) that problems special to the definition of property rights in natural gas (or any other fugacious resources) created difficulties that would not be applicable to coal or timber. Under the "Rule of Capture", prevailing in all states, every owner of land was entitled to extract gas (or oil) by a well bottomed under his parcel, even though the mineral that he extracted originated in a part of a "pool" of gas under another's land. Thus no owner was obliged to consider the full "user cost" of his extraction (*i.e.*, the cost of foregoing the present dis-

[19] Oklahoma v. Kansas Natural Gas Co., 221 U.S. 229, 239-41 n.1 (1911).
[20] 221 U.S. 229 (1911).
[21] 262 U.S. 553 (1923).
[22] *See generally supra* note 3.
[23] 221 U.S. 229, 254 (1911).
[24] *Id.* at 266.

counted value of the revenues that might be derived from later extraction of the mineral). Like two boys attacking a single milk shake with two straws, the owners were likely to extract the mineral at a faster than optimal rate. In any event, the Court would presumably have answered that argument with the response that it gave to the states' vaguer conservation claim: that the negative commerce clause did not permit states to solve the problem by "discrimination" against other states.[25] Oil and gas producing states have indeed now adopted legislation that mitigates the "Rule of Capture" problem in ways that are free of any explicit discrimination (and perhaps free of discriminatory effects): laws restricting the number of wells and the rate of production at each well, and laws facilitating "unitisation" (which, by co-ordinating extraction under a single operator, should effectively cause the pool to be extracted in approximately the way a single owner would choose).[26]

The two cases suggest the long-standing character of the Court's commitment to the principle that express discriminations against interstate commerce violate the commerce clause; at the time they were decided the Court purported to follow a rule that did not directly mention discrimination. The principles then nominally controlling were that all economic activity was inherently either "local" or "interstate"; that only state taxes or regulation on interstate commerce were vulnerable under the negative commerce clause; and that work on a product "before the commencement of its movement from the State" constituted only local commerce, even though the product was later going to embark upon an interstate journey.[27] In the period of the *Oklahoma* and *West Virginia* cases these principles had been applied to sustain *nondiscriminatory* state taxes on the extraction of various minerals.[28] The Court might have invoked them to uphold the Oklahoma and West Virginia statutes, determining, for example, that corporate formation was a "local" act preceding any interstate commerce. Indeed, Justice Holmes took that view. He dissented in *Pennsylvania v. West Virginia* on the ground that "the products of a State until they are actually started to a point outside it may be regulated by the State notwithstanding the commerce clause."[29] But the Court chose to forego that approach, invoking an anti-discrimination principle that was not then part of its official doctrine.

Since the era of *Oklahoma v. Kansas Natural Gas Co.* and *Pennsylvania v. West Virginia*, the Court has expressly embraced the anti-discrimination principle: a state regulation or tax that expressly discriminates against interstate commerce is subject to an overwhelming presumption of invalidity under the negative commerce clauses.[30] This overwhelming presumption is, however,

[25] *Id.* at 262.
[26] For more detail, *see infra* pp. 85-89.
[27] *See, e.g.*, Heisler v. Thomas Colliery Co., 260 U.S. 245, 259-60 (1922).
[28] *Id.*
[29] 262 U.S. 553, 600-01 (1923).
[30] *See, e.g.*, Sporhase v. Nebraska, 458 U.S. 941 (1982).

subject to a substantial exception: where the state *as entrepreneur* pursues a policy of express discrimination against interstate commerce, it may enjoy complete exemption from commerce clause review. Before examining this exception in detail, we look at the way in which the Treaties deal with direct export restrictions.

2. In the European Community

Direct restrictions on imports and exports, as a classic instrument of trade protection, were of course among the primary targets of all three Community Treaties. Each Treaty thus provides for the abolition of any existing quantitative restrictions on trade between the Member States and the prohibition of any new ones.[31] The notion of "quantitative restriction" is perhaps a little broader than that of "quota": it has been defined by an authoritative commentator as

> The totality of rules and administrative measures which wholly or partially exclude the importation of one or several products having regard to their amount or their number, without reference to their quality.[32]

A clear prohibition is, therefore, placed on the use of traditional mechanisms, such as export licensing, through which Member States might seek to restrict the volume of their exports of their natural resources to other parts of the Community in order to ensure the security of their own supplies or to sell into other politically or economically more attractive markets.[33] The fact of joining a common market does not, of course, of itself make these latter objectives irrelevant, and one result of the presence of the simple prohibition on quantitative restrictions is to turn the attention of Member States to ways of achieving similar results without their use as such. Possibilities of doing this under the regime of the ECSC and Euratom Treaties are rather restricted: in the former case by reason of the restricted and highly standardised range of products involved[34] (which makes it hard to restrict imports by the adoption of idiosyncratic national standards), the central powers of the High Authority over pricing,[35] and sweeping basic prohibitions on discrimination in supply, on aids and on special charges;[36] in the latter, by reason of the fairly comprehensive controls exercisable by the Supply Agency over the materials vital to the nuclear industry.[37]

[31] ECSC Treaty art. 4(a); EEC Treaty arts. 30-34; Euratom Treaty art. 93.

[32] Waelbroeck, in 1 J. Mégret, J.-V. Louis, D. Vignes & M. Waelbroeck, Le Droit de la Communauté Economique Européenne 101 (1973), *citing* Fanara, in 1 R. Quadri, R. Monaco, & A. Trabucchi, Trattato istitutivo della Comunità Economica Europea. Commentario CEE 177 (1965) [our translation].

[33] Export licensing may, however, be reintroduced under Community supervision in certain circumstances: *see infra* Ch. III, at pp. 139-40.

[34] *See* ECSC Treaty Annex I, for list and definitions.

[35] ECSC Treaty arts. 61-64.

[36] *Id.* art. 4(b) & (c).

[37] Euratom Treaty ch. VI.

In the EEC, however, with its broad scope and relative absence of central powers, there was an obvious necessity to provide in general terms for the control of quota-like measures; thus the Treaty also prohibits "measures having equivalent effect" to quantitative restrictions. The interpretation of this phrase by the European Court of Justice has been sweeping. Whereas the Commission, in its Directive 70/50[38] (which was concerned only with imports) indicated that the prohibition would only apply in cases where imports were subjected to some rule different from that applied to domestic products, the Court has made it clear that a measure applying in identical terms to domestic products and to imports may likewise be prohibited, if its effect is to disadvantage imported products by reference to domestic ones. Thus a maximum price measure, albeit quite general in its application, might infringe article 30 if "fixed at a level such that the sale of imported products becomes, if not impossible, more difficult than that of domestic products."[39] From this quotation it will appear also that the Court has adopted a very broad concept of "effect". As it said in the case of *Procureur du Roi v. Dassonville*, the Treaty prohibition covers "all trading rules enacted by Member States which are capable of hindering, directly or indirectly, actually *or potentially*, intra-Community trade."[40]

In applying this definition to the facts of the case, the Court showed also that the prohibition could apply not only where there might be an overall reduction in the volume of trade between two Member States, but also where measures were in effect which favoured, within the Community, particular channels of trade (for example, manufacturer-accepted distributors) in preference to others.

This broad judicial approach has been criticised on the ground that the Court, in pursuit of complete freedom of trade between the Member States, may be depriving them of instruments of national economic policy, such as price control, which the Community itself is in no position to wield.[41] The Court has certainly not been blind to this problem,[42] which cannot be resolved simply by resort to the rather narrow grounds of exemption from the rules on quantitative restrictions to be found in article 36. It first indicated a path towards a solution in holding, in *INNO v. Vereniging van de Kleinhandelaars in Tabak (ATAB)*,[43] that measures of equivalent effect to quantitative restrictions do not include measures such as fiscal measures which are specifically referred to elsewhere in the Treaty, nor to those which "are *per se* permitted as being the visible or hidden expression of powers retained by the Member States" (as,

[38] J.O. L 13, 19 Jan. 1970, p. 29, 1970 O.J. (spec. Eng. ed.) 17.

[39] Case 65/75, Tasca, [1976] E.C.R. 291, 308, para. 13.

[40] Case 8/74, Procureur du Roi v. Dassonville, [1974] E.C.R. 837, 852, para. 5 (emphasis supplied).

[41] D. Wyatt & A. Dashwood, The Substantive Law of the EEC 102 (1980).

[42] As witness the remarks of Lord Mackenzie Stuart, one of its members, in his The European Communties and the Rule of Law 76-79, 96-102 (1977).

[43] Case 13/77, [1977] E.C.R. 2115.

for example, powers of general economic management). The path was marked out more clearly in *Rewe-Zentral AG v. Bundesmonopolverwaltung für Branntwein*, the so-called "*Cassis de Dijon*" case.[44] Here the Court held to be contrary to article 30 the German rule applicable alike to imports and to domestic production, forbidding the sale of fruit liqueurs with an alcohol content of less than 25 percent, a rule whose effect was to prevent the importation of authentic Cassis de Dijon, with an alcohol content of 15-20 percent. In reaching this conclusion the Court, without referring to article 36, accepted as legitimate obstacles to movement within the Community resulting from disparities in national marketing laws which were

> necessary in order to satisfy mandatory requirements relating in particular to the effectiveness of fiscal supervision, the protection of public health, the fairness of commercial transactions and the defence of the consumer.[45]

Despite the protestations of the German Government that stiff drinks represent a lesser danger to the public health than do insidiously habit-forming weaker ones, the Court could not find that the 25 percent requirement fell within any of the above categories. It appears though that in the future measures apparently applying indifferently to imports and domestic production will be tested against such broad categories of purpose, rather than the narrower ones listed in article 36.

The extent to which this approach, which combines vigour and flexibility, can be transposed from the field of import restrictions to that of export restrictions is now open to serious doubt, notwithstanding the near-identity of the relevant Treaty articles. Up to 1979 one would have said with some confidence that the Court would apply the same tests to export as to import restrictions, and in particular, that a measure which appeared to apply indifferently to exports and to goods destined for the home market could infringe article 34 if in fact its effect was to make exports more difficult.[46] Now one cannot be so sure. In 1979 a three-judge chamber of the Court decided the somewhat bizarre case of *Groenveld v. Produktschap voor Vee and Vlees*.[47] It was claimed there that a rule forbidding Dutch meat processors to possess or process horsemeat, on the ground that this was the only way of making sure that the horse-loving British and Germans would not reject all Dutch sausages on suspicion of their containing horsemeat, was a measure equivalent to a quantitative

[44] Case 120/78, [1979] E.C.R. 649.

[45] *Id.* at 662, para. 8.

[46] In the United States, such prima facie nondiscriminatory state legislation that may disproportionately burden out-of-state consumers (and may thus have been intended to secure an improper advantage for the state) has been (so far as energy is concerned) largely confined to taxes. United States treatment of that issue is discussed *infra* at pp. 64-77. The problem has also arisen as an offshoot of state efforts to solve common pool problems in oil and gas development, and in that context is considered *infra* at pp. 88-89.

[47] Case 15/79, [1979] E.C.R. 3409.

restriction on Dutch exports of processed horsemeat. This claim the Court rejected, saying that article 34 covers only measures

> which have as their specific object or effect the restriction of patterns of exports and thereby the establishment of a difference in treatment between the domestic trade of a Member State and its export trade in such a way as to provide a particular advantage for national production or for the domestic market of the State in question at the expense of the production or trade of other Member States. This is not so in the case of a prohibition like that in question which is applied objectively to the production of goods of a certain kind without drawing a distinction depending on whether such goods are intended for the national market or for export.[48]

The decision was explicitly affirmed by the full Court in *Oebel*,[49] in which German legal restrictions on night baking and night deliveries from bakeries were alleged to have an effect equivalent to quantitative restrictions on exports for bakeries located near the German border. The Court, following *Groenveld*, rejected this argument on the ground that the restrictions were that of economic and social policy and applied by virtue of objective criteria to all bakery undertakings, without distinguishing between the domestic and export trade of the State. It thus refused to consider trade-restricting effects, despite a strong plea from the Advocate-General that it should not adopt one test for import restrictions and a different, less demanding one, for export restrictions.

Both these cases concerned controls on production, but the language used by the Court was quite general.[50] How does this *Groenveld* formula apply to situations in which Member States appear to be seeking to afford domestic consumers preferential access to national production of oil and gas, along the lines considered by the Supreme Court in *Oklahoma v. Kansas Natural Gas Co.*[51] and *Pennslyvania v. West Virginia*?[52]

In 1973 the Dutch Government, becoming concerned about the depletion of its natural gas reserves and fearing that increased exports of gas would unduly hasten the end of the period of Dutch self-sufficiency in this fuel, announced that no new exports of natural gas by its producing licensees would be permitted unless substantial new finds were made. In pursuance of this policy of reservation of natural gas to national use, the Government blocked the execution of a contract for the sale of gas from the continental shelf by Placid Oil to the German distribution company Ruhrgas. Placid complained to the Commission. It was hard to see how the Government's decision could be other than a measure of equivalent effect to a quantitative restriction on exports, and the Commission, acting under article 169 of the Treaty, sought and ob-

[48] *Id.* at 3415, para. 7.
[49] Case 155/80, [1981] E.C.R. 1993.
[50] For a comparative application of the principle to retail trade, *see* Case 75/81, Blesgen v. Belgium, [1982] E.C.R. 1211.
[51] 221. U.S. 229 (1911).
[52] 262 U.S. 553 (1923).

tained a lifting of the restriction and the withdrawal of the policy.[53] Nonetheless, when the Dutch Government revised its concession terms for offshore production licences in 1976, it provided for the inclusion in future licences of quite explicit provisions which could be used in just the way condemned in 1973. Petroleum and natural gas produced under the licence may not be sold otherwise than under a sales agreement approved by the appropriate Minister. Also, if the Minister decides that gas produced under the licence is needed for supply in the Netherlands, the licensee may be required to supply it to the Dutch gas distribution network, Gasunie.[54]

Gasunie has always been a monopoly buyer of natural gas from the giant landward Groningen field, under the terms of the 1963 concession granted to a consortium of Shell, Esso and the Dutch State Mines. This arrangement was not designed to preclude the export of the gas, and Gasunie has concluded contracts with Belgium, France, Germany and Italy. Here too, however, state control has been tight. Gasunie, as we have already seen, is half owned by the state (as to 10 percent directly, as to 40 percent through the wholly owned Dutch State Mines). We consider its position as a state enterprise in the next section: here we should note that ownership powers were not thought by the Dutch Government to be enough.

Under an agreement of 1963 between Gasunie and the Dutch Government, the Minister of Economic Affairs has power to approve the selling prices and destination of natural gas handled by Gasunie. In 1974, when it became apparent, in the wake of the oil crisis, that the Netherlands was selling so much of its gas cheaply abroad under long-term contracts that it might have to import for home consumption at higher prices, the Government, in addition to the policy of reservation of new supplies referred to earlier, reinforced its powers over prices through the passage of the *Wet Aardgasprijzen*. This Act gives the Minister of Economic Affairs statutory powers to fix minimum prices for the sale of natural gas, in cases where the agreed prices do not properly reflect its market value. If a minimum price decision is not respected, the Minister may order the cutting-off of the relevant supply. These powers apply alike to domestic supply and to exports; they have only been applied in one case, involving a domestic sales contract, but their existence clearly facilitated the success of the exercise of price renegotiation in which Gasunie, together with a high-level Dutch Government representative, engaged with foreign gas buyers in the late 1970s. A number of questions have been asked in the European Parliament about the

[53] *See* The Times (London), 5 Jan. 1973 and 4 May 1973; EC COMMISSION, SEVENTH GENERAL REPORT ON THE ACTIVITIES OF THE EUROPEAN COMMUNITIES (1973), at para. 109 (1974); EIGHTH GENERAL REPORT (1974), at para. 105 (1975).

[54] *See* Besluit van 6 februari 1976, houdende uitvoering van artikel 12 van de Mijnwet continentaal plat [Stb. 1965, 428] ten aanzien van opsporings- en winningsvergunningen voor of mede voor aardolie of aardgas, Stb. 1976, 102, Ch. III, arts. 24-25 [hereinafter Decree of 6 Feb. 1976].

Act and the renegotiations,[55] but no action has been taken in this regard by the Commission, which has taken the view that the Act itself does not contravene Community law, though its application could be contrary to article 34.

It does not seem to us that the Dutch "national reservation" can be redeemed by the more relaxed test for export restrictions propounded in *Groenveld.* It explicitly controls destination of oil or gas, rather than production, and gives the Member State the power to discriminate between destinations. The same is true of the 1963 agreement between the Dutch Government and Gasunie. More problematical, perhaps, is the *Wet Aardgasprijzen.* If a high minimum price were fixed for gas exports, and a lower one — or no minimum at all — for domestic consumption, a specific preferential effect for the domestic market would be easy to show. The task might be harder where a single price were set for both markets, even if this minimum were calculated to reduce gas exports by rendering Dutch gas unattractive in foreign markets. This does not in fact appear to have been the effect of the law and of the negotiations that followed it.

More complex restrictions are to be found in the United Kingdom's offshore oil regime. Within this regime, the principal vehicle of regulation is the terms of the licences, granted by the Secretary of State for Energy, under which companies and consortia enjoy rights to explore for and produce petroleum from the United Kingdom continental shelf.[56] Notwithstanding the fact that the resources being produced are, at least up to the time of production, subject to the proprietary rights of the Crown, and that the licences are in the nature of a contract transferring property, it seems clear that licence controls are capable of falling within articles 30-36, either directly or via article 90(1), which applies the rules of the Treaty to measures enacted by Member States "in the case of ... undertakings to which Member States grant special or exclusive rights." We look at this provision in more detail in the next section, but it certainly seems apt to cover the holders of petroleum production licences. Under the licence, the licensee may not produce any oil or gas otherwise than in accordance with a development programme approved by the Secretary of State for Energy or a production consent given by him.[57] In addition, he is re-

[55] *See, e.g.,* Written Question No. 703/74 (Mr. Gerlach), O.J. C 108, 15 May 1975, p. 13.

[56] Licences are granted under the Petroleum (Production) Act, 1934, 24 & 25 Geo. 5, ch. 36, as applied in relation to the continental shelf by the Continental Shelf Act 1964, ch. 29, and contain standard terms set out from time to time in regulations made under the former Act (currently, the Petroleum (Production) Regulations 1982, S.I. 1982 No. 1000). For texts and commentary *see* UNITED KINGDOM OIL AND GAS LAW (T. Daintith & G.D. Willoughby eds., 2nd ed. 1984 (looseleaf)) and THE LEGAL CHARACTER OF PETROLEUM LICENCES: A COMPARATIVE STUDY ch. 8 (T. Daintith ed. 1981).

[57] *See* clause 14(1) of the Model Clauses for Petroleum Production Licences in Seaward Areas, Petroleum (Production) Regulations 1982, S.I. 1982 No. 1000, Sch. 5.

quired to land any oil or gas he produces in the United Kingdom, unless the Secretary of State consents to an alternative landing place.[58]

On the United Kingdom's accession, this latter provision attracted the immediate attention of the Commission, as a possible restriction on exports. Though it has never initiated any complaint on this score or published its views, its reasoning is not hard to reconstruct. The landing requirement could arguably be regarded as a straightforward quantitative restriction on exports, *if* direct transportation of United Kingdom continental shelf crude oil and gas production to other countries is regarded as an export from the United Kingdom. Under the Community's origin rules, the origin of continental shelf production is deemed to be that of the coastal state to which the shelf appertains.[59] This general principle would suggest that we would find an "export" in the above situation; but these origin rules have never been applied to crude oil, natural gas and other petroleum products.[60] Moreover under United Kingdom law, the continental shelf is clearly not part of the customs territory; for the purposes of the import duties regime products of the continental shelf coming to the United Kingdom are merely "deemed" not to be imported by express legislative provision.[61] Provision also exists for the making of governmental orders adopting origin rules for continental shelf production (of the United Kingdom itself and of other countries) of similar effect to the general Community rule;[62] but no such orders have been made. On the other hand, the United Kingdom does, as a matter of practice, include direct shipments of oil from the continental shelf to third-country ports in the crude oil exports it reports to the Commission under Regulation 388/75 "on notifying the Commission of exports of crude oil and natural gas to third countries."[63]

At the least, these considerations leave room for some doubt as to whether such direct transportation constitutes export. If not, the landing requirement arguably operates to produce an indirect effect on eventual exports of United Kingdom continental shelf production in the crude state. Where the offshore producer would otherwise have the choice between direct shipment to a foreign port and direct shipment to the United Kingdom, the requirement that he land his production in the United Kingdom adds extra transport and transshipment costs to the export of oil and gas in its crude state. It is calculated, in fact, to render such an option less attractive than the refining of the crude in the United Kingdom and its export as higher-value petroleum products. Seen in

[58] *Id.* cl. 27.

[59] Council Regulation (EEC) 802/68, art. 4(2)(h), J.O. L 148, 28 June 1968, p. 1, at p. 2, 1968 O.J. (spec. Eng. ed.) 165, 166.

[60] *Id.* art. 3 and Annex I. For an unsuccessful Commission attempt to fill this lacuna, *see* its proposal for a Council Regulation on the common definition of the concept of the origin of petroleum products, O.J. C 124, 15 Oct. 1974, p. 1.

[61] Customs and Excise Duties (General Reliefs) Act 1979, ch. 3, § 14(2) (*re-enacting* Finance Act 1967, ch. 54, § 2(1)).

[62] *Id.* § 14(3) (*re-enacting* Finance Act 1967, § 2(2)).

[63] O.J. L 45, 19 Feb. 1975, p. 1.

this light, the provision thus offers an excellent example of the rule which applies both to production destined for home or export use but which in fact only burdens the latter.

If the first of these approaches is adopted, and the landing requirement is viewed as a direct restriction on exports, then the *Groenveld* test, again, gives no comfort to the United Kingdom Government. If, on the other hand, we take the second approach, and see the requirement as discriminatory *in effect* as between producers with home and overseas refineries, the *Groenveld* test would appear to apply, with the consequent need to show a "specific object or effect"[64] of a discriminatory nature. This should not be too difficult. The production restrictions in question in *Groenveld* and *Oebel* had purposes — in the one case the restriction of the use of horsemeat, in the other employee welfare — which would stand regardless of the conditions of international trade. This is not the case with the landing requirement. If there were no interstate trade in crude oil, it would have no reason to exist.[65] Its specific effects are, therefore, to be related to trade, and should be seen as including the discrimination we have identified.

If these restrictions do potentially constitute *prima facie* obstacles to trade, our next task must be to see if they can be justified. This question has two aspects: The first — legal, hypothetical — is what would the European Court of Justice decide if these restrictions were alleged to breach article 34? The second — factual — is how can we explain their survival and the acquiescence of the Commission?

As yet there are no grounds in European Court of Justice jurisprudence to suggest that measures which do specifically affect patterns of export can be justified otherwise than on the basis of article 36.[66] The United Kingdom Government is understood to have sought to justify the landing requirement by reference to the need to ensure the effective assessment and collection of taxes due from holders of production licences, but this objective is not among those mentioned in article 36 as providing grounds for derogating from the ordinary prohibition on quantitative restrictions. Even if it were, one might expect the Court to hold that the particular restriction imposed was excessive having regard to the objective pursued.[67] More appropriate, in the context of article 36, is the concept of public security. Secure energy supplies are surely a key element in the economic life of any nation; moreover a drastic reduction in supply levels might jeopardise other interests referred to in article 36 — public

[64] *See supra* text accompanying notes 47-48.

[65] It is not clear if the Court in *Groenveld (supra* note 47) thought that the restriction in question there was explicable on grounds other than those related to Dutch sausage exports. The existence of such grounds was certainly arguable.

[66] In Case 113/80, Commission v. Ireland, [1981] E.C.R. 1625, the Court explicitly held that the *Cassis de Dijon* approach would not operate where different rules applied to domestic production and to imports.

[67] As in Case 104/75, de Peijper, [1976] E.C.R. 613.

order, and human life and health.[68] This public security ground was successfully invoked by the Irish Government before the Court in the *Campus Oil* case,[69] as a justification of its rule that importers of oil products must buy a proportion of their requirements from the only (state-owned) Irish oil refinery. Without these compulsory purchases, it was said, the refinery would have to close and Ireland would be completely dependent on supplies of oil products from abroad, which would jeopardise public security. The Court held that while the requirement was contrary to article 30, it could be excused under article 36 if the compulsory purchases were the minimum necessary to ensure the continued operation of the refinery. While this decision deals with imports rather than with exports, it warrants mention here because it had earlier appeared that resort to article 36 might be excluded in relation to security of oil supply because the Community had itself taken measures with this objective, in particular, a scheme modelled on the International Energy Programme for sharing available oil supplies among members countries in times of supply difficulties.[70] The Court has in the past held that Member States cannot rely on article 36 to justify restrictions on trade, such as border inspections, when the matter is covered by a harmonising directive under article 100.[71] It now appears that the Court will look to see if the Community measure provides *adequate protection* for the Member State interest in question.[72] Even with this qualification there appears no scope for justifying *export* restrictions by reference to supply difficulties for the purpose of article 36, because here there are Community measures addressed to this precise problem and envisaging the introduction of such restrictions after an appropriate Community procedure. Decisions under article 103(4), to be discussed in detail later, envisage the suspension of export licences for crude oil and petroleum products by Member States, in prescribed circumstances, with the authorisation and under the supervision of the Council and Commission.

This being so, we are deprived of one immediate explanation for Commission acquiescence in the continuation of these restrictions. The likelihood of their successful defence before the European Court of Justice appears low.

[68] *See*, for the UK, the Royal Proclamation of 12 Dec. 1973, S.I. 1973, vol. III, p. 8039, referring to "the reduction of oil supplies reaching Great Britain" as one of several events "calculated ... to deprive the community ... of the essentials of life" and thus justifying the declaration of a state of emergency under the Emergency Powers Act, 1920, 10 & 11 Geo. 5, ch. 55.

[69] Case 72/83, Campus Oil Ltd. v. Minister for Industry and Energy, [1984] 3 C.M.L.R. 544.

[70] The Programme is to be found at 14 I.L.M. 789 (1975), and is discussed in Claudy, *The International Energy Agency*, 14 NAT. RES. LAW. 454 (1982). *See also infra* Ch. III, at pp. 137 & 140-41.

[71] *See* Case 137/77, City of Frankfurt-am-Main v. Firma Max Neumann, [1978] E.C.R. 1623; Case 138/77, Firma Herman Ludwig v. Free and Hanseatic City of Hamburg, [1978] E.C.R. 1645.

[72] Case 72/83, *Campus Oil*, [1984] 3 C.M.L.R. 544, 568, para. 28.

The reason for their survival must be sought in non-legal considerations. In the absence of discussion by the Commission, one can only speculate as to the reasons for its abstention from enforcement. It is clear that enforcement would be politically costly for the Commission, given the generally nationalistic attitude of all Member States to the natural resources found on their territories or in their jurisdiction, an attitude fully shared by the United Kingdom and the Netherlands; but to this political calculation there may perhaps be added a strategic one. In the absence of a common policy for exports of crude oil, gas and petroleum products from the Community — the commodities have since 1957 remained "for the present" outside the general Community regime of export liberalisation[73] — the British and the Dutch controls represent the only instrument available within the Community to secure that exports of the major oil and gas resources existing within the Community are directed primarily to the Community rather than to third countries.[74] In the absence of an appropriate Community regime, therefore, these powers may be better than nothing, and the Commission may monitor their exercise to ensure that (as largely appears to have been the case up to the present time) their use in fact is not incompatible with the objectives of the Community energy policy.

Such political and strategic considerations are much less likely to influence the actions of private parties who might be affected by the British and Dutch restrictions. We simply record here that none of these restrictions has been the subject of privately instituted litigation, postponing until a later part of the book our discussion of why this might be so.

B. The State as Market Participant

We have already seen that in the European Communities, state enterprises are heavily involved in the production, transformation and distribution of energy.[75] This is not the case in the United States, where outside the sphere of electricity utilities there is little state ownership in the energy sector. A consequence of the central place of state enterprises within the range of instruments of industrial policy in the European Community countries is that the problems they create for free-trade are explicitly (if not always unambiguously) confronted by the Treaties. The United States Supreme Court, on the oth-

[73] *See* Council Regulations (EEC): 2603/69, art. 10, J.O. L 324, 27 Dec. 1969, p. 25, at p. 28, 1969 O.J. (spec. Eng. ed.) 590, 593; 1934/82, art. 1, O.J. L 211, 20 July 1982, p. 1. *See also infra* Ch. III, at pp. 136-37. This exclusion has recently been confirmed by the European Court of Justice which was called upon to interpret the above regulations in their application to UK oil export policy in Case 174/84, Bulk Oil (Zug) AG v. Sun International Trading Co. Ltd., (Judgment of 18 Feb. 1986, not yet reported).

[74] The UK also exports oil to other members of the International Energy Agency (especially to the US) and to Finland.

[75] *See supra* Ch. I, at pp. 10-15.

er hand, has had to find its own way in dealing with an unfamiliar phenomenon. Despite the relatively underdeveloped character of United States law in this area, we discuss it first, since the need for the courts to proceed from first principles has caused them, and commentators on their decisions, to spell out in detail the problems which the provisions of the European Community Treaties assume to exist and to resolve.

1. In the United States

(a) The Basic Doctrine

In two recent cases the Supreme Court has treated express discriminations in favour of state residents or in-state businesses as beyond judicial scrutiny under the negative commerce clause. One case involved the state as a provider of cash subsidies for local environmental clean-up; the other, the state as owner and operator of a cement plant. Although the Court alludes to the exception as being for the state as market participant, the scope of the exception is not clear and may fall short of what that label suggests.

Hughes v. Alexandria Scrap Corp.[76] considered a statute by which the Maryland legislature, distressed at the aesthetic effects of derelict automobiles on the Maryland landscape, sought to accelerate the process of recycling these eyesores into steel. It used a stick (in the form of a recurring fine for wreckers who retained any such auto for more than one year) and a carrot, in the form of a bounty to scrappers for each vehicle scrapped. The goal and effect of the bounty were to increase the economic incentives to engage in vehicle scrapping and in supplying derelicts to the scrappers. (The latter effect occurred because economic forces[77] led the scrappers to share the bounties with wreckers and others who supplied them with derelicts.)

The statute also sought to stimulate the recycling process by reducing the risk that owners, claiming they had not abandoned their vehicles, might sue reprocessors for conversion. It specified various title documents which the scrapper might obtain from his supplier and which (1) were made a condition of the scrapper's receiving the bounty and (2) would clear his title against suits by persons claiming to be the vehicle's owner. But, recognising that these title documentation needs might themselves impede the recycling process, the legislature set up a looser regime for "hulks" (defined as vehicles over eight years old and without an engine or otherwise totally inoperable). As originally enacted in 1967, the statute altogether exempted hulks from the provision that title documents should be a condition of a scrapper's receiving the bounty. But a 1974 amendment tightened the hulk provisions, and, for the first time, introduced a distinction in the treatment of in-state and out-of-state scrappers. An in-state scrapper could receive the bounty on a hulk merely by furnishing to the state an agreement by his supplier to indemnify him for any third-party

[76] 426 U.S. 794 (1976).
[77] *Id.* at 797.

claims arising from its distruction. But an out-of-state processor was required to produce the same documentation for hulks as for all other vehicles.[78]

As the statute explicitly discriminated against out-of-state scrappers, a conventional application of the standard test would have required the state to meet a quite heavy burden of justification. Maryland sought to meet that burden by an argument that the distinction was appropriate in order "to reduce the amount of state funds paid for destruction of Maryland-titled hulks abandoned in the States where those processors are located instead of in Maryland." The special documentary burden on out-of-state processors was said to further the legislature's purpose that the subsidy go only to removal of Maryland-titled hulks *abandoned in Maryland.*[79]

The majority, however, held that the case could be resolved in Maryland's favour without assessing that explanation. In quite broad terms, it appeared to hold that when a state "has entered into the market itself to bid up [the] price" of a good or service,[80] and restricts its purchases to its own citizens or businesses with the state, the negative commerce clause does not "require independent justification for such action."[81]

The crucial distinction appeared to be between two roles of the state: that of a *regulator of the market* and that of a *market participant.* While acknowledging that as a result of the discriminatory feature of the Maryland statute, hulks "will tend to be processed inside the State rather than flowing to foreign processors",[82] the Court said:

> But no trade barrier of the type forbidden by the Commerce Clause ... impedes their movement out of State. They remain in Maryland in response to market forces, including that exerted by money from the State.[83]

Reeves, Inc. v. Stake[84] applied that principle or a related one to allow South Dakota, as the owner of a cement plant, to meet a sudden increase in the scarcity of cement by adoption of a sales policy expressly discriminating against out-of-state purchasers. The policy adopted was to supply "all South Dakota's customers first and to honor all contract commitments, with the remain-

[78] An exception to the documentation requirements allowed any scrapper, even out-of-state, to rely on a mere "Wrecker's Certificate" as to hulks supplied by licensed wreckers. *See id.* at 801.

[79] *Id.* at 804-05.

[80] *Id.* at 806.

[81] *Id.* at 809-10.

[82] *Id.* As noted above, Maryland's concern was not that provision of the subsidy to out-of-state processors on equal terms would accelerate the flow of hulks out of state, but that it would increase the risk of the subsidy's being used to scrap Maryland-titled vehicles abandoned outside Maryland.

[83] *Id.* at 809-10.

[84] 447 U.S. 429 (1980).

ing volume allocated on a first come, first served basis."[85] The *Reeves* majority interpreted *Hughes* to distinguish "between States as market participants and States as market regulators",[86] and set out to justify the distinction. First, it invoked "the long recognized right of trader or manufacturer, engaged in an entirely private business, freely to exercise his own independent discretion as to parties with whom he will deal."[87] Second, it invoked what may be termed a "bittersweet" argument: that the state's exposure to some of the burdens of private business when it acts in a proprietary sense entitles it to some of the freedoms such firms enjoy. Its exposure to some of the bitter entitles it to enjoy some of the sweet.[88] Third, it argued that "the competing considerations in cases involving state proprietary action often will be subtle, complex, politically charged, and difficult to assess under traditional Commerce Clause analysis."[89]

The justifications articulated by the Court do not seem very persuasive. The first one raises, but does not answer, the question whether, for negative commerce clause purposes, it is appropriate to treat the state in this situation as similar to a private enterprise. The second justification, alluding to greater burdens imposed on the state when it acts in a proprietary sense, referred to that portion of the opinion in *National League of Cities v. Usery*[90] in which the Court had asserted that Congress had far greater leeway under the commerce clause to regulate state proprietary activities than to regulate certain decisions that a state made in carrying out its non-proprietary functions (there the wage rates it might offer to its employees). The assertion in *National League of Cities* represented the Court's conclusion that the Tenth Amendment's protection of the states from Congress' affirmative power under the commerce clause was weaker as to states' proprietary activities than as to their non-proprietary ones. Conventional reasoning by analogy, therefore, would suggest that the states' protection from the negative commerce clause should similarly be weaker as to their proprietary activities. The marvel of the "bittersweet" argument is that, by invoking it, a court can at will reverse the results to which garden-variety legal reasoning would lead.

Finally, the last argument does not seem to distinguish state proprietary activities from non-proprietary ones. Are not the competing considerations in the latter also "subtle, complex, politically charged, and difficult to assess"?

85 *Id.* at 432-33. One should perhaps say that the Court seemed to apply the *Hughes* principle, for Justice Powell, author of the majority opinion in *Hughes*, dissented in *Reeves* and argues that the *Reeves* majority had got *Hughes* all wrong: *id.* at 447-54.

86 *Id.* at 436.

87 *Id.* at 438-39.

88 *Id.* at 439.

89 *Id.*

90 426 U.S. 833, 854 (1976), *overruled*, Garcia v. San Antonio Metropolitan Transit Auth., 105 S. Ct. 1005 (1985).

Hughes and *Reeves* together, then, seem to establish a substantial degree of exemption from the negative commerce clause for state decisions as to the use of state funds or property, though *Reeves* may limit that exemption to cases where the funds or property are being used in a proprietary way. They do not establish any powerful rationale for the exemption, nor do they offer many clues as to its scope.

(b) Possible Rationales

Professors Wells and Hellerstein have suggested that discriminatory action in the use of state property contributes just as much to the balkanisation of the economy — which the negative commerce clause seeks to avoid — as does discriminatory state regulation. Therefore, they argue, the justification for the *Hughes-Reeves* exemption must be some countervailing value. They find such a value in the state's interest in its fiscal autonomy.[91]

The first point — that balkanisation is balkanisation whatever its origins — is of course correct. But the Court might nonetheless justify the *Hughes-Reeves* exemption, in terms of the negative commerce clause itself, by arguing that state discrimination in the use of its property is subject to more effective inherent limitations than state regulatory discrimination. If so, perhaps there is less need for judicial scrutiny.

The core of the inherent limitation is that the state government will often bear a direct and significant portion of the costs of the discrimination, just as would a private business so foolish as to discriminate against out-of-state customers or suppliers. These costs may take the form of either direct impacts upon the state's budget (in the form of increased expenses or of revenues foregone), partial frustration of the state's programme, or both. For example, South Dakota could have responded to the increased scarcity of cement by raising its prices to market-clearing levels and enjoying the resulting profits. Its decision to hold prices down prevented the market from clearing, and required some non-price system for allocation. A significant cost of its election to allocate discriminatorily, therefore, was its loss of the revenues that would have flowed from adjusting its prices up to market-clearing levels.[92]

[91] Wells & Hellerstein, *The Governmental-Proprietary Distinction in Constitutional Law,* 66 Va. L. Rev. 1073, 1125-35 (1980).

[92] The majority opinion in *Reeves,* therefore, is wide of the mark when it argues that a decision against the state "would rob South Dakota of the intended benefit of its foresight, risk, and industry". 447 U.S. at 446. In truth, a decision against South Dakota would merely have required it to enjoy those benefits in the *form* of extra revenues for its cement plant, rather than in the form of preferential access to cement for South Dakota users.

If one viewed South Dakota's *sole* purpose in establishing its cement business as the protection of its citizens from cement shortages (*see id.* at 430) and if one reads the "shortage" concept loosely to include any increase in scarcity which entails sharp price increases in order to clear the market, then the Court's observation would be correct. The Court itself, however, noted that South Dakota's venture into the ce-

In *Hughes* the price paid by the state for its discrimination is less clear. If the restraints imposed on non-Maryland scrappers were arbitrary in terms of the purposes of the programme (as we may assume under the majority's language of complete exemption), then it would seem probable that, for any given level of subsidy, Maryland's discrimination forced it to achieve less scrapping. (It forewent at least some scrapping services of non-Maryland scrappers who by hypothesis could scrap more efficiently than Maryland ones who instead enjoyed the business.) Alternatively, Maryland's achievement of any particular scrapping goal (say, 100,000 hulks per year), would require higher levels of subsidy, since the arbitrary burden on out-of-state scrappers would reduce the quantity of scrapping services drawn into the market for each subsidy dollar. Consumers who arbitrarily refuse to deal with any set of potential suppliers ordinarily must pay some price for their arbitrariness, and there is no reason to believe that Maryland, as a purchaser of scrapping services, would be any exception.

There are two significant limits to the above argument. First, the political process may characteristically reflect the value of local gains from discrimination against outsiders more fully than those from non-discrimination; thus, even when the local community would gain (in aggregate utility or wealth) from non-discrimination, a discriminatory policy may emerge from the political arena. Second, states pay a price for regulatory discrimination too, so that the distinction is far from absolute.

The first limit is most apparent when we consider state subsidies of education — with respect to which, the Court has said, the state may discriminate in favour of its citizens.[93] If one formulates the state's goal in purely utilitarian terms — generating the most education for any given level of expenditure — then it is apparent that discrimination against out-of-state students would compromise the goal. But, given American population mobility, a state bears only a portion of the cost of failing to educate a promising non-citizen who is excluded because the state prefers a less promising citizen. More important, the political process in all likelihood impresses upon state legislators only the palest reflection of the state's cost. Outsiders who would be attracted to a state by its nondiscriminatory provision of education, and whose later presence within the state might generate widespread benefits, have no voice in its political process. Even in *Reeves*, the political pressure imposed by South Dakota cement users seeking preferential access to cement probably is more effective

ment business was based in part on a perception of the "substantial profits" being enjoyed by the private producers of cement (*id.* at 431 n.1); it seems probable that that perception entailed an idea that South Dakota's entry into the business would enable its government to enjoy comparable profits.

[93] *See* Starns v. Malkerson, 401 U.S. 985 (1971), *aff'g*, 326 F. Supp. 234 (D. Minn. 1970). The opinion is seemingly approved in the later case of Vlandis v. Kline, 412 U.S. 441, 452 n.9 (1973). *See* discussion below on the current status of the case and its reasoning.

than that of South Dakota taxpayers who perceive that the state's discrimination imposes serious opportunity costs on the state treasury.[94]

The second limitation — the existence of parallel self-restraints for state regulatory discrimination — is pervasive. In the two cases that parallel *Reeves*, for example, invoking explicit state restrictions on the export of natural gas (*Oklahoma v. Kansas Natural Gas Co.*[95] and *Pennsylvania v. West Virginia*[96]) the artificial restriction of the market presumably reduced revenues for in-state natural gas producers (forcing them to forego the price that would clear the market if out-of-state purchasers had been free to compete for natural gas supplies with in-state ones). Parallel in-state costs can be found for every discriminatory state regulation.

Nonetheless, the Supreme Court might with some plausibility hypothesise that state expenditure of funds is subject to more effective inherent limitations against discrimination against out-of-staters than is state regulatory action. State legislators and officials actually doling out state funds in voluntary market-like exchanges seem somewhat more likely to perceive the costs of their discriminations — more likely to conform to the profit-maximising goals of private owners similarly limited to voluntary exchange transactions — than when they wield the state's regulatory authority.

The suggestion by Wells and Hellerstein that the state's interest in fiscal autonomy accounts for the different treatment of discriminatory use of property raises a further question: why should that interest be greater than the state's interest in its regulatory autonomy? Here surely a key element is the likelihood that the citizens preferred by the state's discrimination have contributed in taxes towards the accumulation of the property involved. The equitable claims deriving from those tax contributions have great force.[97] Unfortunately, how-

[94] Effective political pressure in the above comparison should be measured as a relative matter — in proportion, say, to the aggregate utility or wealth losses of the cement users and taxpayers, respectively.

[95] 221 U.S. 229 (1911).

[96] 262 U.S. 553 (1923).

[97] *See* Varat, *State "Citizenship" and Interstate Equality*, 48 U. Chi. L. Rev. 487, 523, 526-30 (1981). As noted below, even Justice Brennan, the justice most hostile to allowing states to differentiate against nonresidents on the basis of the prior contribution rationale, is willing to give it weight in some circumstances.

Within the EEC, under secondary legislation (Council Regulation (EEC) 1612/68, J.O. L 257, 19 Oct. 1968, p. 2, 1968 O.J. (spec. Eng. ed.) 475), the Community worker enjoys all rights afforded to national workers in matters of public housing, including eligibility to be entered on housing lists on equal terms (art. 9). Children of Community workers are granted the right to education and vocational training, including education grants for higher education (art. 12). For a comparison of American and European law in this field, *see generally* Garth, *Migrant Workers and Rights of Mobility in the European Community and the United States: A Study of Law, Community and Citizenship in the Welfare State*, in I/3 Integration Through Law 85 (1986).

ever, the status of the prior contribution rationale is in grave doubt. As recently as 1971, the Supreme Court seemed to accord it some weight by its per curiam affirmance of *Starns v. Malkerson*,[98] upholding the University of Minnesota's tuition preferences for residents against an Equal Protection challenge. The University charged nonresidents about double the price for residents, and provided that no student was eligible for resident classification "unless he has been a bona fide domiciliary of the state for at least a year immediately prior thereto."[99] Minnesota justified the rule as an attempt "to achieve partial cost equalization between those who have and those who have not recently contributed to the State's economy through employment, tax payments and expenditures therein."[100] The lower court accepted that theory, and the Supreme Court affirmed per curiam.

Other cases, however, before and after *Starns*, indicate a rejection of the theory. In 1969, in *Shapiro v. Thompson*,[101] the Court sustained an Equal Protection challenge to state provisions denying welfare payments to residents who had been domiciliaries for less than one year. The state invoked the contribution rationale, but the Court rejected it on the grounds that such "reasoning would logically permit the State to bar new residents from schools, parks, and libraries or deprive them of police and fire protection."[102]

Post-*Starns* cases occasionally treat the contribution rationale with as much or more disdain. *Vlandis v. Kline*[103] invalidated a Connecticut tuition preference scheme, under which an unmarried student was conclusively restricted to nonresident status if his legal address was outside Connecticut for any part of the one-year period immediately prior to his application for admission, and a married student was conclusively deemed nonresident if his legal address was outside Connecticut at the time of his application. The *Vlandis* Court simultaneously affirmed the continued authority of *Starns*,[104] disapproved the contribution rationale, and asserted that Connecticut's scheme for classification of students as resident or nonresident was only arbitrarily related to that rationale, assuming it to be valid.[105]

Judicial prohibition of all preferences for in-state citizens would sharply reduce the political viability of state subsidy programmes. The Court's relatively accommodating treatment of such discrimination in the context of higher edu-

98 401 U.S. 985 (1971), *aff'g*, 326 F. Supp. 234 (D. Minn 1970).

99 326 F. Supp. at 235-36.

100 *Id.* at 240.

101 394 U.S. 618 (1969).

102 *Id.* at 632.

103 412 U.S. 441 (1973).

104 *Id.* at 452 n.9.

105 *Id.* at 450 n.6. In addition, in Memorial Hospital v. Maricopa County, 415 U.S. 250 (1974), the Court invalidated a statute that conditioned free access to non-emergency medical care on prior domicile in the county for one year. The state's effort to invoke the contribution rationale met with the pronouncement, "We rejected this contributory rationale both in *Shapiro* and in *Vlandis v. Kline.*" *Id.* at 266.

cation may then simply arise from an implicit decision that such activities are valuable and ought not to be discouraged.

On this analysis, the Court's decision in *Zobel v. Williams*,[106] invalidating Alaska's scheme for making residents' shares of a fiscal dividend proportional to years of residency in Alaska since statehood, could be seen as merely reflecting a view that fiscal dividends do not deserve comparable protection.[107] But in *Zobel* the Court spoke much more broadly, characterising the contribution rationale as "not a legitimate state purpose"[108] at all. The decision represents exceptional hostility to the contribution rationale, for the Court purports to evaluate Alaska's law by its most deferential standard (where it reviews only for "minimum rationality").[109] *Zobel* expressly notes that the Supreme Court's action in *Starns* was only a summary affirmance, which "is not to be read as an adoption of the reasoning under review."[110] *Zobel* also suggests that the residency requirement upheld in *Starns* was considered as "a test of *bona fide* residency, not a return on prior contributions to the commonwealth."[111] It seems at least curious that the Court should now banish state consideration of probable prior contributions to outer darkness, yet allow a state to treat nonresidents very differently from residents, when the difference in probable prior contribution is such a salient distinction between the two.

It is hard to accept *Zobel*'s scornful treatment of the contribution rationale at face value. As suggested above, it seems to deprive the states' power to prefer residents of at least one of its most substantial bases. And even Justice Brennan, the Court's most resolute opponent of any special treatment for state discrimination in the use of its funds, appears to acknowledge that a state could justify lower fees for its citizens, in access to some state resources, by reference to the citizens' tax contributions to the maintenance of those resources.[112]

As a result of the Court's vacillating disapproval of the fiscal autonomy-prior contribution argument, and the inadequacies of its own contentions in *Reeves*, the true rationale for *Hughes* and *Reeves* remains in doubt. A consequence of this uncertainty is to cast grave doubt upon the Court's pronouncements that state discriminations affected by the doctrine are utterly free of fed-

106 457 U.S. 55 (1982).

107 This analysis leaves uncertain the rationale for the Court's rejection of discrimination in provision of medical care (Memorial Hospital v. Maricopa County, 415 U.S. 250 (1974)) and in provision of welfare (Shapiro v. Thompson, 394 U.S. 618 (1969)).

108 457 U.S. 55, 63 (1982).

109 *Id.* at 60-61.

110 *Id.* at 64 n.13.

111 *Id.*

112 Baldwin v. Montana Fish & Game Comm'n, 436 U.S. 371, 401 (1978) (Brennan, J., dissenting). Justice Brennan, however, would apparently permit the fee differential only to the extent that it precisely compensated for citizens' contribution by taxes. *See id.*

eral judicial review under the negative commerce clause. We turn below to some circumstances under which the Court may whittle the fact of state ownership down to a mere factor to be "weighed" in its review process.

(c) Possible Bases for Not Applying the Doctrine to Energy Resource Development

The *Reeves* opinion appears to suggest that the fact that the state was only exercising its property rights would have less exempting effect if the item owned were a natural resource. The district court, finding against South Dakota, had invoked a traditional commerce clause argument — that if one state could "hoard" resources originating within its borders, then other states could do the same, with the result that "embargo may be retaliated by embargo and commerce would be halted at state lines."[113]

The Supreme Court's response was to find that "Cement is not a natural resource, like coal, timber, wild game or minerals." [114] Rather, it said, cement is "the end product of a complex process whereby a costly physical plant and human labor act on raw materials."[115] (Mineral extractors may be startled at the Court's apparent assumption that minerals attain usable commercial form without "a costly physical plant and human labor".) The Court's response implictly assumes that a state has *less* freedom to restrict commerce in natural resources than commerce in other goods. It is hard to find a basis for such a view in the Court's prior negative commerce clause decisions. To take an obvious example, several of the Court's most aggressive interventions have involved milk, which is not conventionally listed as a natural resource.[116] And several of the Court's opinions have at least suggested that because of a state's vital interest in its natural resources, it was entitled to *more* leeway with respect to them, rather than less.[117]

The hint in *Reeves* may prove significant, however. In an extensive treatment of "State 'Citizenship' and Interstate Equality",[118] Varat suggests a number of reasons why courts might hesitate to extend the *Hughes-Reeves* immunity to discriminatory state conduct in the development of its energy resources — either as a passive lessor or an active entrepreneur. As to nonreproducible natural resources, he argues that (1) the state is likely to have a monopoly power that would render discriminatory behaviour more offensive; and that (2) at

113 447 U.S. 429, 443 (1980).

114 *Id.*

115 *Id.* at 444.

116 Baldwin v. G. A. F. Seelig, Inc., 294 U.S. 511 (1935); Dean Milk Co. v. Madison, 340 U.S. 349 (1951); H. P. Hood & Sons v. Du Mond, 336 U.S. 525 (1949).

117 Baldwin v. Montana Fish & Game Comm'n, 436 U.S. 371 (1978); Hughes v. Oklahoma, 441 U.S. 322 (1979). *See also* Hudson County Water Co. v. McCarter, 209 U.S. 349 (1908); *cf.* New England Power Co. v. New Hampshire, 455 U.S. 331, 338-39 n.6 (1982).

118 *See* Varat, *supra* note 97.

least as to natural resources to which the state acquired title with its original sovereignty, the contribution rationale is inapplicable.[119]

Neither point seems conclusive. A state's "monopoly" power with respect to a resource turns on the ease or difficulty faced by buyers who seek to replace that resource with substitutes. This does not correlate perfectly — or perhaps even very well — with the item's being either "natural" or "nonreproducible". If the state of Pennsylvania owned and exploited its oil and gas resources, for example, it would surely exercise no monopoly power, given Pennsylvania's trivial share of United States and global production. By contrast, South Dakota appears to have enjoyed dramatic regional monopoly powers in the cement industry: when it adopted its preference for residents, Reeves, Inc., the complaining purchaser, was compelled to reduce its production (of ready-mix concrete) by 75 percent.[120]

Conceivably, however, the Court might identify nonreproducible natural resources as an area with sufficient risk of monopoly power that as a categorical matter it should be denied the advantages of the *Hughes-Reeves* doctrine. One might find support for such a principle in the Court's persistent commerce clause scrutiny of the operation of state-owned highways, even in opinions explicitly acknowledging the fact of state ownership.[121] These operations might be seen to entail such a high potential for the exercise of monopoly power through the obstruction of natural routes that judicial intervention is appropriate.[122] (Alternatively, one might set such cases aside as special on the ground that they deal with *arteries* of commerce, or that the state's activity is govern-

[119] *Id.* at 556-57

[120] 447 U.S. 429, 452-53 n.4 (1980) (Powell, J., dissenting).

[121] Buck v. Kuykendall, 267 U.S. 307, 314 (1925); South Carolina State Highway Dep't v. Barnwell Bros., Inc., 303 U.S. 177, 187 (1938).

[122] South-Central Timber Dev., Inc. v. LeResche, 511 F. Supp. 139 (D. Alaska 1981), invalidated an Alaska requirement that purchasers of state timber do "primary manufacture" of the timber within Alaska, on the basis of the negative commerce clause. It refused to apply *Reeves* on the ground that a natural resource "is not a commodity which, when needed, is capable of being readily produced by any state at any time." *Id.* at 142. Western Oil & Gas Ass'n v. Cory, 726 F.2d 1340 (9th Cir. 1984), ***aff'd without opinion by an equally divided Court,*** 105 S. Ct. 1859 (1985), invalidated part of a state's charges for transmission of petroleum products from plaintiffs' coastal processing plants to mainland markets over state lands, on the basis of the negative commerce clause, as well as the Constitution's prohibitions on state duties on imports or exports (art. I, § 10, cl. 2). The Court distinguished *Reeves* on the bases that the state (a) owned the relevant lands in its sovereign rather than in its proprietary capacity and (b) had a complete monoply over the sites used for rights-of-way. 726 F.2d at 1343. (It is true that once plaintiffs located their plants where they did, a relation of bilateral monopoly existed between them and the state as owner of the coastal barrier strip. But the state *seems* only to have been enforcing pricing regulations in effect at the time of plaintiffs' *choosing* the location; depending upon the uniqueness of the offshore sites, there may or may not have been a competitive market for the rights-of-way at the time of that choice.)

mental rather than proprietary, if we take as authoritative *Reeves*'s apparent limitation of the rule to proprietary activities as authoritative.)

The state's acquisition of the resource as part of its sovereignty may weaken the contribution rationale, but it surely does not obliterate it. So long as the state refrains from outright sale of the resource *in situ*, it incurs opportunity costs: foregoing the revenue that such sale would generate. Residents may be said to have contributed in the sense of having denied themselves the chance to enjoy some combination of lower taxes or superior services which they could have obtained by having the state sell the resource. And if a state may sell off its natural resources for top dollar, invest the proceeds in some enterprise having nothing to do with natural resources, and then discriminate in the operation of the latter, it is not clear why it should not be free to discriminate in development of the natural resource itself.

Another variation that Varat suggests is the difference between (1) a state's discriminatory distribution of a resource (say, a two-tiered auction, in which a resident's bid would enjoy a 50 percent premium when measured against a nonresident's, so that the former's bid of $100 would beat the latter's bid of $149) and (2) its conditioning access to the resource on the recipient's engaging in discrimination against nonresidents. The state's exercise of its leverage in the second case, he suggests, begins to look like regulation.[123]

Whatever their force, one or more of these elements may account for the Court's decision in *Hicklin v. Orbeck*[124] in 1978 (after *Hughes* and before *Reeves*). The Court there overturned a statute by means of which Alaska used its ownership of oil-and-gas properties to obtain hiring preferences for its citizens. The statute required all oil-and-gas leases, easements or unitisation agreements to which the state was a party to include provisions requiring the other parties to hire citizens of Alaska in preference to non-citizens, no matter how much more qualified the latter might be. The Court unanimously found the statute to violate the Privileges and Immunities Clause of article IV, section 2. The Court expressly rejected the relevance of *McCready v. Virginia*,[125] an 1876 decision upholding against Privileges and Immunities Clause attack a Virginia statute prohibiting non-citizens from planting oysters in navigable streams. It argued that "the connection of the State's oil and gas with much of the covered activity is sufficiently attenuated so that it cannot justifiably be the basis for requiring private employers to discriminate against non-residents."[126]

[123] *See* Varat, *supra* note 97, at 560-64.

[124] 437 U.S. 518 (1978).

[125] 94 U.S. 391 (1876).

[126] 437 U.S. at 529. In South-Central Timber Dev., Inc. v. Wunnicke, 104 S. Ct. 2237 (1984), the Court invalidated, on negative commerce clause grounds, a comparable provision imposing, as a matter of contract, requirements of domestic (*i.e.*, in-Alaska) processing of timber purchased from the state. The Court distinguished *Reeves* on the grounds that in *South-Central Timber* a "natural resource" was in-

Clearly Alaska was seeking to collect the economic rents on its mineral resources in a special form — the contract promises of those engaged in exploitation of the resource. Presumably bidders for Alaskan oil and gas depressed their bids to reflect the costs of compliance with the hiring preference, and the revenue thus foregone measures the cost to Alaska of obtaining the preference it would have secured (if the contracts had been enforced). If Alaska had collected that portion of the rents on its mineral resources in cash — as an owner is normally free to do — it could probably have used that sum to finance employment for its residents and possibly other types of preferential job programmes.[127] A priori, there is no reason to think that it achieved more discriminatory preference per dollar (paid or foregone, as the case may be) by one device than by the other. Nonetheless, the Court invalidated this direct approach.

In sum, the likelihood of exceptions to the *Hughes-Reeves* immunity seems great. All three of the elements suggested by Varat — the natural and nonreproducible character of energy resources, their acquisition by the state as an aspect of sovereignty, and the use of economic rent to induce discrimination by private persons — may play a role in the evolution of these exceptions.[128]

2. In the European Community

Here, the economic importance of state enterprises clearly poses a problem in relation to the construction of a common market involving the free movement of natural resources among the Member States. They may be used by Member States as an instrument of control whose working is far less transparent than

volved and that the restriction burdened resale, as well as that there was an effect on foreign commerce. 104 S. Ct. at 2245.

Justice Rehnquist, in dissent, suggested various ways, similar to those referred to in the text immediately below, by which Alaska could achieve the same effect by devices that do fall within the market-participant exception. 104 S. Ct. at 2248-49.

127 *See* Varat, *supra* note 97, at 546-48. In White v. Massachusetts Council of Constr. Employees, Inc., 460 U.S. 204 (1983), the Supreme Court upheld an executive order of the Mayor of Boston requiring that the work force on any project financed in whole or in part with city funds be composed at least 50 percent of bona fide residents of the city.

128 The Court has recently established a proposition that may facilitate a retreat from *Hughes* and *Reeves* in natural resource and other issues. In United Building & Constr. Trades Council v. Mayor & Council of Camden, 465 U.S. 208 (1984), the Court held that the market participant exception was not applicable when discriminatory state legislation was challenged under the Privileges and Immunity Clause. The Court reiterated, however, its observation in *Hicklin v. Orbeck* that state ownership of the property concerned "is a factor — ... often the crucial factor — to be considered in evaluating whether the statute's discrimination against noncitizens violates the Clause." *Id.* at 221, *quoting* Hicklin v. Orbeck, 437 U.S. 518, 529 (1978). Under the Privileges and Immunities Clause, however, the challenger must establish that the state law burdens interests "'fundamental' to the promotion of interstate harmony". 465 U.S. at 218.

that of the restrictions we have already discussed. This possibility is not confined to the natural resources sector: it exists in relation to all aspects of intra-Community trade. The Framers of the Treaties were well aware of it, and made specific provision for it. We shall look briefly at the general organisation of the relevant Treaty provisions, then at their application in relation to the disposition of oil and gas from the United Kingdom sector of the North Sea.

All three Treaties start from the position that they do not seek to modify or prejudice the system of (property) ownership applicable in the various Member States.[129] The relevant provisions have been interpreted, by authoritative commentators, to mean that Member States remain free to change the nature of property in particular enterprises, as by nationalisation; they do not, however, have the effect of rendering other provisions of the Treaties inapplicable.[130] It remains necessary, therefore, to consider how far the general provisions of the Treaties apply in relation to public enterprises. In the ECSC and Euratom Treaties, no explicit provision is made for public enterprises, and it may therefore be assumed the ordinary rules apply save where their terminology makes them plainly inappropriate.[131] In the EEC Treaty, on the other hand, special rules appear in articles 37 and 90.

The first of these articles, appearing in the Title on free movement of goods, makes provision for the adjustment of state monopolies of a commercial character. It was designed to cope with the existence, in the original Member States, of import monopolies like the French tobacco monopoly, whose activities would otherwise have needed to be summarily terminated by reason of their clearly restrictive effect on imports. Instead it provided that the structure and operation of such monopolies should gradually be adjusted so that no discrimination would exist between nationals of Member States regarding the conditions under which goods are procured and marketed. In addition, the introduction of new measures contrary to this prohibition on discrimination or restricting the effect of the other free movement of goods provisions was prohibited. While the definition of "state commercial monopoly" is loosely drafted and is possibly broad enough to embrace BNOC and BGC, there is little interest in discussing their activities in this context. Any period during which a monopoly, existing at the date of the Treaty itself or of the Treaty of Accession, could benefit from a regime of "adjustment" more indulgent than that of

129 ECSC Treaty art. 83; EEC Treaty art. 222; Euratom Treaty art. 91. This last article is qualified by reference to the fact that the Treaty does affect property ownership, vesting in the Community the right of ownership of all special fissile materials produced in or imported into the Community and subject to its safeguards system (art. 86). The practical significance of this arrangement has been minimal.

130 *See, e.g.*, Minervini, in 2 R. Quadri, R. Monaco & A. Trabucchi, Trattato istitutivo della Comunità europea del carbone e dell'acciaio. Commentario CECA 1199-1202 (1970).

131 For example, the provisions of ECSC Treaty art. 66 referring to "transactions ... bringing about ... a concentration between undertakings" cannot be applied literally to concentrations arising from nationalisation legislation.

the general Treaty provisions is now over;[132] nor may the creation of a new state monopoly be used as a means of temporarily avoiding the operation of the free movement of goods provisions of the Treaty. In consequence the significant question in relation to bodies like BNOC and BGC in the United Kingdom, or Gasunie in the Netherlands, is not whether they qualify as monopolies in terms of article 37, but whether, as state enterprises, they are subject to the general competition provisions of the Treaty, and whether they can be used by the Member State as a vehicle for securing effects which, if sought through regulatory mechanisms, would fall foul of Treaty provisions such as articles 16 and 34.

These issues are directly addressed by article 90. It comprehensively recognises the special position and potentialities of public enterprises (and, indeed, of private enterprises in a special relationship with the state). Article 90 (1) regulates the behaviour of the state towards public undertakings and undertakings to which it grants special or exclusive rights, providing that the state shall not, in their case, enact or maintain in force any measure contrary to the rules contained in the Treaty, especially its general prohibition against discrimination, and its rules on restrictive agreements, abuse of dominant positions, and state aids. The aim of this provision is to prevent the Member State from using its *de facto* or *de jure* powers over such undertakings to secure behaviour by them which would be contrary to the Treaty, or which, while not explicitly forbidden to the public enterprise, would be a breach of the Treaty if engaged in directly by the Member State itself.[133] The Commission, indeed, has gone further, and argues that it imposes a positive obligation on the Member State to intervene to correct any such behaviour by these enterprises.[134] Article 90(3) gives the Commission a particularly powerful weapon of intervention for the enforcement of the obligation created by article 90(1), enabling it to address directives or decisions to Member States without the formalities normally requisite under article 169 where Member States are alleged to have failed in their obligations under the Treaty.[135]

[132] For the original Member States the transition period expired on 31 Dec. 1969; for the Member States acceding in 1972 (who all said that at that time they had no such monopolies) it expired on 31 Dec. 1977.

[133] Examples of such behaviour would be discrimination against customers and suppliers on grounds of nationality (art. 7), or the operation of measures of equivalent effect to restrictions on imports or exports (arts. 30, 34) as by confining sales in times of shortage to domestic buyers. Such conduct is directly forbidden only to monopolies (art. 37) and enterprises with a dominant position (art. 86), and even then only when certain other conditions are satisfied.

[134] EC COMMISSION, SIXTH REPORT ON COMPETITION POLICY (1976), at para. 274 (1977).

[135] The first such directive, Commission Directive (EEC) 80/723, O.J. L 195, 29 July 1980, p. 35, was designed to secure transparency in financial relationships between Member States and certain classes of state enterprises. This Directive does not apply to state enterprises in the energy sector, though these arrangements may be extended

Article 90(2) has a quite different purpose. In relation to a narrower category of undertakings than is covered by article 90(1), that is, "undertakings entrusted with the operation of services of general economic interest or having the character of a revenue-producing monopoly", it provides for the application of Treaty rules insofar as this "does not obstruct the performance, in law or in fact, of the particular tasks assigned to them". Even this limited exception is restricted by the requirement that "the development of trade must not be affected to such an extent as would be contrary to the interests of the Community". The Court has moreover held that since this provision derogates from general Community law, the class of undertakings which can benefit from it must be strictly defined.[136]

The compatibility of the position and activity of BNOC, BGC and Gasunie with the Treaty as applied by article 90 is not free from doubt. The situation of each of these bodies is somewhat different: the complexity of the operation of Treaty rules in these cases may perhaps best be judged by looking closely at one case, that of BNOC.[137]

One of the objectives of the Labour Government, in creating BNOC in 1975,[138] was to increase its control not only over the production, but also over the use and disposition, of North Sea oil. As we have already noted,[139] in common with the majority of other Western states with oil resources, the vehicle chosen by the United Kingdom for the exploitation of the offshore oil over which the Geneva Convention granted it "sovereign rights to explore and exploit"[140] was a concession system under which companies (or more normally,

to such enterprises in the future. A challenge to the legality of the Directive has been rejected by the European Court. Joined Cases 188-190/80, French Republic, Italian Republic and the United Kingdom v. Commission, [1982] E.C.R. 2545.

136 Case 127/73, BRT v. SABAM and NV Fonior, [1974] E.C.R. 51 & 313.

137 Evans, *United Kingdom North Sea Oil Policy and EEC Law,* 7 Eur. L. Rev. 355-68 (1982). As this chapter was being revised for the press, the UK Government announced its intention to abolish BNOC and to suspend the operation of the participation arrangements under which BNOC had traded such large quantities of North Sea oil. Participation may be reactivated in times of supply emergency, and a limited amount of oil trading (essentially in royalty oil taken in kind) will continue to be done on the Government's behalf by a new body, the Government Oil and Pipelines Agency, but this will clearly not possess the economic importance of BNOC nor pose the same problems for EEC law. For details *see* United Kingdom Department of Energy, Press Notice No. 29, 13 Mar. 1985; Oil and Pipelines Act 1985, ch. 62. While the discussion that follows is now, therefore, of mainly contingent or historical interest, it seems worth retaining, since broadly similar problems may arise, or have arisen, in relation to other major energy enterprises such as DONG in Denmark or BGC in the UK (now to be transformed into a single private enterprise by legislation introduced into Parliament on 28 Nov. 1985).

138 Petroleum and Submarine Pipe-Lines Act 1975, ch. 74, Pt. I.

139 *See supra* Ch. I. at pp. 10-14.

140 Geneva Convention on the Continental Shelf, 1958, art. 2(1), 499 U.N.T.S. 311, 52 A.J.I.L. 834 (1958).

consortia of companies) received licences, contractual in nature, to explore for and produce oil in a given area, the oil becoming, on production, the property of the licensees in return for the payment of a royalty to the state. Despite the increasing elaboration of the controls provided for in the licence, the Government felt it needed a degree of involvement in the process of production and disposition of oil that only the creation of public enterprise could give. To secure for BNOC, as its primary chosen vehicle, rights under future production licences was easy; to do the same in relation to existing licences, which in 1975 already covered the most prospective United Kingdom continental shelf territory and often had some forty years to run, was more difficult, and required individual agreements between BNOC and all licensees having interests in commercial fields. The process of making these agreements went under the title of "voluntary participation" but involved the exercise of considerable governmental pressure: a government undertaking that acquiescing licensees would be financially neither better nor worse off as a result of conceding participation, the withholding of necessary governmental consents under existing licences until agreement was reached, and threats of discrimination against recalcitrants in the award of future licences.

Between 1975 and 1979, therefore, BNOC acquired two kinds of participating interests in North Sea licences. In new licences and, through some individual deals, in some old ones, it acquired equity interests, under which it contributed its percentage of exploration and production costs (usually 51 percent) and became entitled, without further payment, to an equivalent percentage of oil produced. In existing licences, through "voluntary participation", it obtained more limited rights as described in the following paragraph.

The Conservative Government that succeeded to power in 1979 has had very different objectives for BNOC, seeing its essential function as that of obtaining for the nation secure access to a substantial share of United Kingdom continental shelf oil, and designing licence terms so as to give BNOC rights in future licences similar to those acquired through "voluntary participation". In pursuance of this policy, equity interests acquired by BNOC in the period 1975-1979 have been hived-off into a new company, Britoil, 51 percent of whose shares have been sold to the public, the remainder being retained, for the present, as a direct government shareholding. In consequence, the activities and functions of BNOC under pre- and post-1975 licences are now essentially similar. In each case, a commercial find under a licence leads (or has led) to a situation in which BNOC becomes a joint holder of the licence with the existing licensees, but does not make any contribution to the costs of operations under the licence nor receive any of the economic benefits thereof. Under the ordinary forms of consortium agreement used in the North Sea, the licensees, who hold the whole licence jointly without any division into shares, agree on the apportionment among themselves of the oil produced, ordinarily in proportion to their contribution to the costs of exploration and exploitation. The participation agreements or licence terms provide that BNOC becomes a signatory of these consortium agreements with the right to take, at its election, up to 51 percent of the oil that would otherwise have accrued to each of the original partners, on paying the market price therefor. At first sight this looks like an option to purchase. BNOC and the United Kingdom Government maintain, however, that it is not: when, they say, BNOC exercises its right of election, it does no more than claim for itself oil which, by reason of its being a holder of the licence, already belongs to it jointly along with the other licensees.

The way in which BNOC comes into possession of at least 51 percent of crude oil produced on the United Kingdom continental shelf calls for two comments from the standpoint of Community law.

First, the argument that BNOC gets the oil as producer rather than as purchaser, while not developed within the context of a consideration of Community law[141] nonetheless seems relevant to it, in the sense that it may be easier under the Community legal regime to accept BNOC as the holder of production rights to 51 percent of United Kingdom oil production than as the imposed purchaser of that percentage from the producers. A straightforward legislative measure compelling the sale of 51 percent of produced oil to BNOC would, it is suggested, be treated as a measure equivalent to a quantitative restriction on exports, in that it deprives the producers of all possibility of themselves exporting any of that 51 percent. In *Pigs Marketing Board v. Redmond*[142] a sale obligation of this kind, operating in relation to an agricultural product covered by a common organisation of the market, was held incompatible with articles 30 and 34 and with the market regulation; it is not clear that the Court would have reached the same decision in the absence of a Common Market organisation, but the Advocate-General clearly would have done so.[143]

Second, conditions for the award of licences clearly place BNOC in a position quite different from that of other prospective licensees. The conditions assure BNOC's presence on every licence which covers an oil-producing field; at the same time, they deny BNOC the possibility of obtaining any of the economic rent from the field, by stipulating that it is to acquire its 51 percent of production by payment of the market price therefor to the other members of the consortium. The differences in roles and objectives between BNOC and other licensees are so fundamental that it seems inappropriate to speak of "discrimination" between BNOC and them in the process of awarding licences, still less of discrimination on grounds of nationality such as is forbidden generally under article 7 and, with specific reference to the grant of oil production li-

[141] The primary purpose of the arrangements referred to, and of the characterisation of the rights of BNOC, was to provide BNOC with a legal remedy whereby, in case licensees should renege on their undertaking to deliver oil, BNOC would be able to insist on the delivery of that oil (rather than being restricted to a remedy in damages). Labour Ministers were also enabled to say with conviction that they had succeeded in restoring to the nation property in the greater part of North Sea oil production.

[142] Case 83/78, [1978] E.C.R. 2347.

[143] *Id.* at 2385-86.

cences, by Council Directive 64/428.[144] It is in any event implicit in the wording of article 90(1) itself that Member States retain the power to grant "special or exclusive rights" to undertakings. Thus the existence of the Italian state television monopoly was approved by the European Court in *Sacchi.*[145] *A fortiori*, it must be open to the Member State to devise arrangements whereby a public enterprise obtains rights to only a majority share of production of a particular good, particularly, perhaps, where the State is the owner of the good in its "unproduced" form, that is, as oil *in situ*, and could if it wished undertake all exploration and production operations itself.[146] If that majority share causes BNOC to possess a dominant position on the relevant market, of course, then behaviour which would, under article 86, constitute an abuse of that position must be avoided, a point that we look at in a moment.

If the mode of endowing BNOC with oil steers clear of Treaty obstacles, its use of that oil is still fully subject to applicable rules of the Treaty. Article 90(2) can hardly apply to relieve BNOC of any of its obligations.[147] Oil supplies are certainly of general economic interest, but it is doubtful if the trading of oil in the crude state can be qualified as a "service"; moreover, it is not easy to see how any of the tasks assigned to BNOC by its constitutive statute[148] could be obstructed by any of the Treaty rules. Through participation agreements, shares in licences granted since its formation, and otherwise, BNOC is presently able to dispose of over half of North Sea oil production. This suggests the possibility of a dominant position within the meaning of the Treaty, a possibility which is reinforced by the fact that it is BNOC that sets the market price of North Sea crude: its prices are followed by other licensees in the sale of their remaining production. BNOC, as the world's biggest trader of light, low-sulphur crude,[149] could possibly be dominant even in a market including other such crudes from Nigeria, Libya and Algeria, although it has been persuasively argued that BNOC's actions are too constrained — by OPEC pricing policies on the one hand and by the inflexibility of its contractual commitments to acquire United Kingdom oil on the other — to permit it to exercise dominance.[150] If BNOC discriminates in its sales policy, for example

[144] Of 7 July, 1964, 1964 J.O. 1871 (23 July 1964), 1963-1964 O.J. (spec. Eng. ed.) 151.

[145] Case 155/73, [1974] E.C.R. 409, 430.

[146] As done, in fact, by countries such as Mexico, Venezuela, and the USSR.

[147] For the relevant wording of art. 90(2) *see supra* text accompanying note 136. The arguments in the remainder of this paragraph do not necessarily apply in relation to BGC and Gasunie where gas distribution functions may more easily attract the labels of services of general economic interest.

[148] Petroleum and Submarine Pipe-Lines Act 1975, ch. 74, § 2(1) & (2).

[149] *World Glut Tests the Mettle of BNOC's Oil Traders,* Financial Times, 15 Dec. 1982, at 9.

[150] Roeber, *The Formation of North Sea Prices,* in RECENT DEVELOPMENTS IN UK PETROLEUM LAW 1982-83, at 87-89 (T.C. Daintith ed. 1983). This view is borne out by the losses BNOC sustained in the oil market in 1984-85, when it was buying oil at

by refusing supply to the European market or making sales on more onerous terms, it may thus have a defence against an allegation of abuse of a dominant position contrary to article 86. However, the former kind of discrimination, and perhaps the latter also, would seem at the same time to entail a breach by the United Kingdom Government of articles 34 and 90(1), as being an indirect mode of maintaining a measure equivalent to a quantitative restriction on export, if it could be shown that the Government had required or authorised, or perhaps simply not prevented,[151] such behaviour. The existence of such a breach would not depend on BNOC's holding a dominant position in terms of article 86.

There is no evidence that BNOC has discriminated, at least against other Member States, in its supply policy. In a supply crisis, satisfactory criteria on which to demonstrate the existence of discrimination might be difficult to find, as the case of *BP v. Commission*,[152] arising out of the 1973-1974 oil crisis, clearly demonstrated. Behaviour of companies in such crises (and particularly of companies dealing in crude oil) is now likely to be determined more by the operation of the emergency regimes set up both within the EC[153] and in the wider framework of the International Energy Programme of the OECD,[154] so that BNOC's position will in practice probably create little difficulty. Another national objective which may be furthered by its supply dominance is that of the support of the United Kingdom refining industry by the discouragement of exports in crude form. Here, while there is no reason to impeach BNOC's behaviour, the scheme of participation agreements appears to be protectionist in character. Some agreements with licensees who have refining facilities in the United Kingdom (but none with others) provide for BNOC to sell back to the original licensees the oil it has taken from them under the participation agreements. The assumption is that the licensee will refine it in the United Kingdom. The Secretary of State is, however, empowered to order BNOC not to sell back some or all of this oil. No special circumstances for the exercise of this power are indicated, but there is a clear understanding that a change by the licensee in its practice of refining in the United Kingdom may call it into play. Despite the somewhat eleborate way in which the arrangements have been designed, it would not seem unduly difficult to find in such a Ministerial order to BNOC a measure of equivalent effect to a quantitative restriction on the export of the quantities of crude in question.

term prices higher than those it could obtain for sales on the spot market. Its term prices were not lowered for fear of provoking a collapse of the world crude market, of offending OPEC, or both. It was these difficulties which led the Government to decide upon the abolition of BNOC and of this state trading function: *see supra* note 137.

151 *See supra* text accompanying note 134.

152 Case 77/77, [1978] E.C.R. 1513.

153 *See infra* Ch. III, at pp. 138-42.

154 *See infra* Ch. III, note 181.

These comments on the creation, position and behaviour of BNOC have been designed to exemplify, in a case of particular economic significance, the additional issues in the application of the Treaty rules to the natural resources sector which arise when the state intervenes actively through the agency of a public body. They have, we hope, shown how questions as to the free movement of goods present themselves in a more complex form, and how questions as to the compatibility of the behaviour of such bodies with the competition rules of the Treaty also arise.

C. Taxation of Energy Sources

1. In the United States

(a) Introduction

Taxation represents a promising device by which a state might try to capture producer or consumer surplus in mineral resources. For example, if a state knew the extraction cost for each unit, it could tax the difference between that cost and the market price (*i.e.*, the economic rent or producer surplus) without having an adverse impact on production.[155] Since the information necessary for such surgical precision is not available at reasonable cost, states will in practice apply cruder measures. But where the economic rents are very large, even the cruder devices may raise very substantial revenue without very serious impacts on production.

Any tax on economic rents may give rise to regional antagonisms. A mineral-rich state that can finance a significant portion of its government expenses out of such rents is in a position to levy general taxes lower than those of less favourably situated states. Alaska, for example, is evidently able to finance 90 percent of its expenditures out of petroleum taxes and royalties, and has eliminated its income tax.[156] This ability, of course, may afford the energy-rich states a substantial edge in the interstate competition for industry and growth.[157] If the owners of the resource are out-of-staters, the tax has the additional advantage (from the taxing state's viewpoint) that its burden is "exported".

[155] However, as pointed out in note 17 *supra,* a tax capturing precisely the economic rent would destroy the market mechanism for allocating a resource over time.

[156] *See* Commonwealth Edison Co. v. Montana, 453 U.S. 609, 646 n.11 (1981) (Blackmun, J., dissenting).

[157] H.H. Landsberg & J.M. Dukert, High Energy Costs: Uneven, Unfair, Unavoidable? (1981), make the point that advantages in access to energy have always played a role in differentials in regional growth. In the days of sail, for example, proximity to lakes or seacoasts provided superior access to the major form of transportation energy of that era, wind and water. Early US mill towns tapped another region-specific energy resource, falling water. *Id.* at 46.

Recently, tax exportation in a different sense — through alleged imposition of the economic burden of the tax on out-of-state *consumers* — has been a subject of acute controversy. Suppose that all of the specific mineral resource were located in one state and that the extraction industry were highly competitive, but that there were no good substitutes for the mineral at prices close to its market price. Although some or all extracting firms would enjoy economic rents, the competitive character of the industry would preclude any monopoly profits. The happy state in which these firms were located, however, could obtain monopoly profits by taxing extraction of the mineral. The unavailability of good substitutes at comparable prices would mean that the price elasticity of demand for the state's production was very low, with the result that a large share of the tax would be passed on to buyers. To the extent that such buyers were out of state, the tax would be effectively exported, even though in form not discriminatory against out-of-staters.

A recent case, *Commonwealth Edison Co. v. Montana*,[158] appears to confirm the authority of states to impose mineral taxes that have both these effects. State taxation of energy resources is probably not restricted by the negative commerce clause so long as it is neither (1) expressly discriminatory against out-of-state purchasers nor (2) so structured as to compel the inference that the state deliberately sought to discriminate against such purchasers.

The tax challenged in *Commonwealth Edison* was a severance tax on extraction of coal, amounting to about 30 percent of the value at the minehead of almost all Montana coal. According to the pleadings, about 90 percent of the coal was shipped out of state, and evidently all, or nearly all, of the exported coal was subject to long-term contracts under which tax increases were simply added to the purchase price.[159] The Montana legislature's conference committee and at least one of the tax bill's sponsors had, in pressing for its adoption, argued that the tax would be paid largely by nonresidents.[160] The Supreme Court nonetheless rejected claims that the tax infringed the negative commerce clause.

A preliminary step was the Court's disapproval of *Heisler v. Thomas Colliery Co.*,[161] a 1922 decision that had analysed a similar coal tax on the basis of the then prevailing idea that the negative commerce clause could invalidate only taxes that applied to "interstate" commerce.[162] The *Heisler* Court sustained the

[158] 453 U.S. 609 (1981). The case is treated at length in Williams, *Severance Taxes and Federalism: The Role of the Supreme Court in Preserving a National Common Market for Energy Supplies*, 53 Colo. L. Rev. 281 (1982), and the issue of tax exportation is treated at length in Hellerstein, *Constitutional Limitations on State Tax Exportation*, 1982 ABF Research J. 1.

[159] So plaintiffs had alleged, and, since the trial court had dismissed the complaint for failure to state a claim, Montana procedural law required that the allegations be accepted as true. 453 U.S. at 639-42.

[160] *Id.* at 639-41.

[161] 260 U. S. 245 (1922).

[162] *See supra* text accompanying note 27.

tax in question on the ground that the event taxed, severance, occurred before the coal embarked upon its interstate journey.[163] As the Court had dropped that analysis from every other aspect of its negative commerce clause jurisprudence,[164] it proceeded in *Commonwealth Edison* to do so for mineral taxation.

Next came the task of applying the Court's current test for state taxation challenged under the negative commerce clause, that of *Complete Auto Transit, Inc. v. Brady*.[165] The Court there said it would sustain a tax if it met four criteria: "(1) the tax is applied to an activity with a substantial nexus with the taxing State, (2) is fairly apportioned, (3) does not discriminate against interstate commerce, and (4) is fairly related to the services provided by the State."[166]

Plaintiffs argued that the Montana tax violated the third and fourth criteria. But their theory made the asserted violation of both criteria depend upon the *rate* of tax; they conceded the constitutionality of a tax in the range of 12.5 percent to 15 percent.[167] Thus the discrimination was said to arise out of the alleged fact that "the tax was adopted because it would shift an *extraordinary* burden to other States."[168]

The majority, however, considered the "discrimination" and "fair relation" issues separately. Justice Marshall, writing for the Court, dispatched the "discrimination" claim curtly. First, he pointed out that the tax was "computed at the same rate regardless of the final destination of the coal."[169] Second, to answer plaintiffs' claim that a tax was discriminatory for commerce clause purposes "if the tax burden is borne primarily by out-of-state consumers,"[170] he briefly revived *Heisler*. He noted that there plaintiffs had asserted discrimi-

[163] In light of the Court's invalidation, in the same era, of express state barriers against export (*see supra* text accompanying notes 18-30), query whether the Court even then would have applied this doctrine to uphold a statute that taxed coal for export at a higher rate than coal for in-state consumption.

[164] *See, e.g.*, Wickard v. Filburn, 317 U.S. 111 (1942) (congressional power under the commerce clause); Pike v. Bruce Church, Inc., 397 U.S. 137 (1970) (invalidation of state regulations under the negative commerce clause); Complete Auto Transit, Inc. v. Brady, 430 U.S. 274 (1977) (invalidation of state taxes under the negative commerce clause).

[165] 430 U.S. 274 (1977).

[166] *Id.* at 279.

[167] 453 U.S. 609, 620 n.10 (1981).

[168] Brief for Appellants at 33 (emphasis added).

[169] 453 U.S. at 618.

[170] *Id.* This was not a precise or even a fair characterisation of plaintiffs' claim, which in fact was that the tax was discriminatory because it "was adopted because it would shift an extraordinary burden to other states." Brief for Appellants at 33. Thus they were not claiming that a tax was "discriminatory" merely by virtue of its burden being exported, but by virtue of the enacting legislature's intention to achieve that effect. Plaintiffs did not, however, develop an express view as to what proof would be necessary for a judicial finding of such discriminatory intent. *See infra* text accompanying notes 185-95, for discussion of that issue.

nation against interstate commerce on the grounds that Pennsylvania had a monopoly of anthracite coal and shipped 80 percent of the coal out of state. But the *Heisler* court had dismissed those elements as "adventitious considerations".[171]

Thus the majority seemed to interpret *Complete Auto*'s "discrimination" criterion as prohibiting only express discrimination. This was odd, as the Court has employed a broader concept of discrimination in other recent negative commerce clause cases.[172]

As for the claim of plaintiffs that the tax was not "fairly related to the services provided by the State", the majority said that that criterion had nothing to do with the amount or rate of tax in relation to (a) the cost or value of the services actually provided by the state or (b) the costs imposed upon the state by the taxpayer's activity. It meant only "that the *measure* of the tax must be reasonably related to the extent of the contact."[173] The Court found such a relation in the fact that here the measure was the value at the minehead. The Court offered no example of a tax that would fail the reformulated "fair relationship" test.[174] In fact, it is hard to imagine a tax that would do so without also violating the requirement of adequate "apportionment".[175] In any event, the Court evidently intends to let the substance of the "fair relation" test wither away.

We will briefly consider below the discrimination, "fair relation" and apportionment principles as applied to state energy taxation.

171 453 U.S. at 618.

172 *See, e.g.*, Hunt v. Washington State Apple Advertising Comm'n, 432 U.S. 333 (1977), discussed *infra* text accompanying note 187.

173 453 U.S. 609, 626 (1981).

174 The Court was a little less than candid in its implication that it had never suggested that this criterion had anything to do with amounts or values. *Id.* at 625 n.15. In Department of Revenue v. Association of Washington Stevedoring Cos., 415 U.S. 734 (1978), the Court upheld a challenged tax, in part on the basis that nothing *in the record* had suggested that the tax was not fairly related to services and protection provided by the state. It explained that because of the tax-payers' mistaken theory, they had "developed *no factual basis* on which to declare the Washington tax unconstitutional." *Id.* at 750-51 (emphasis added). But the record clearly showed that the tax was to be measured by gross receipts. *Id.* at 738 n.4. If the Court in *Washington Stevedoring* actually intended the "fair relation" test to mean as little as the majority in *Commonwealth Edison* seemed to think, then its suggestion that a better record might have altered the outcome was deceptive. It surely invited litigants and courts to develop factual records, presumably of the sort that plaintiffs would have tried to establish in *Commonwealth Edison* had they reached the stage of trial.

175 *See* Han Rees' Sons v. North Carolina, *ex rel.* Maxwell, 283 U.S. 123, 135 (1931) (tax liability may not be "out of all proportion" to the business transacted by taxpayer in the taxing state).

(b) Discrimination: The Concept of "Exporting" a Tax

A tax whose burden is "exported" beyond the taxing jurisdiction raises special problems. First, to the extent that the burden falls on persons not represented in the taxing state, the ordinary political constraints upon excessive taxation are reduced, as the Supreme Court and many commentators have often observed.[176] This relaxation or skewing of the political constraints on taxation is likely to result in more numerous taxes, taxes pinpointed on activities that ought not to be pinpointed for special taxation, and higher tax rates. This additional tax burden, in turn, will obstruct commerce to some degree. Taxation without representation, thus, may not only be tyranny, but may also chill productive activity.[177] Since these obstructions of commerce clearly arise out of the multi-jurisdictional character of the United States, one may well see them as problems the Framers sought to control when they granted Congress power over interstate commerce.

Second, as the benefits of the tax are presumably enjoyed almost entirely by the citizenry of the taxing state, one could, to the extent that *any* of the burden is exported, perceive a discriminatory *effect*. Thus one might view any such tax as violating the anti-discrimination criterion of *Complete Auto*.

But judicial intervention based upon these insights would soon encounter intractable factual issues, murky line-drawing problems, and a need to invade policy issues normally thought of as belonging to state legislators rather than federal judges. The incidence of a tax turns upon the elasticities of supply and demand for the taxed commodity. As one economist has succinctly put it, "Inelastic suppliers or demanders pay taxes."[178] Thus, if the consumers of Montana coal had no substitutes available at comparable prices — because, say, of Montana's share of the total coal market, special attributes of its coal, or its proximity to such consumers — they would tend to bear a high proportion of the tax. While if the suppliers' readiness to produce was unaffected by the tax,

[176] *See, e.g.*, McCulloch v. Maryland, 17 U.S. (4 Wheat.) 316, 428 (1819); United States v. County of Fresno, 429 U.S. 452, 458-59, 463 n.11 (1977); McGoldrick v. Berwind-White Coal Mining Co., 309 U.S. 33, 45 n.2 (1940); Hellerstein, *State Taxation and the Supreme Court: Toward a More Unified Approach to Constitutional Adjudication?*, 75 Mich. L. Rev. 1426, 1448-50 (1977); Browde & DuMars, *State Taxation of Natural Resource Extraction and the Commerce Clause: Federalism's Modern Frontier*, 60 Ore. L. Rev. 7, 36 (1981); R. Posner, Economic Analysis of Law 509-10 (2d ed. 1977).

[177] Of course, taxes typically chill the activity taxed. Legislatures in which the ultimate payors are under-represented, however, are more likely (when compared with a legislature in which all burdened and benefitted parties are fully represented) to enact a tax whose chilling effect exceeds the benefits deriving from the government's use of the funds. For purposes of this comparison, the "benefits" from the government's use of the funds means (a) the direct benefit from such use, minus (b) the benefits that would have accrued from the use of those funds by the persons who would have received them but for the tax.

[178] D. McCloskey, The Applied Theory of Price 324 (1982).

presumably because a large proportion of the pre-tax price was economic rent, the suppliers would bear most of it.[179] Thus, Montana's export of 90 percent of its coal by no means demonstrated that it was able to export 90 percent of the tax. Determination of price elasticities is complex; a hint of the difficulties is found in the estimates of the long-term price elasticity of demand for motor gasoline, a much studied matter; they range from -0.22 (very inelastic) to -1.3 (quite elastic).[180] Moreover, tax exportation is a matter of degree.[181] What percentage of the tax would have to be exported before the tax either fell as discriminatory, or became subject, as the dissenters argued, to a requirement of special justification?

Thus there are good reasons for rejecting any rule under which courts would find a tax discriminatory on the basis of its being exported. Even a rule which then allowed the state to justify an exported tax by proof of special costs associated with the taxed enterprise[182] would thrust the courts deep into the policy-making duties of state legislatures. In this respect, it would be like rules equating disproportionate impact with constitutionally objectionable discrimination.[183] Such rules tend to sweep very broadly indeed.[184]

179 For a more complete treatment of the incidence problem, *see* Williams, *supra* note 158, at 291-95. If it was true, as claimed, that the US owned 75 percent of Montana coal (Brief of Appellants at 9), then insofar as the tax fell on economic rent it would fall on the US (putting aside the extent to which coal leases made producers effective owners of a portion of that 75 percent).

180 *See* Ford Foundation Study Group, Energy: The Next Twenty Years 90 (1979). *See generally* D. R. Bohi, Analyzing Demand Behavior: A Study of Energy Elasticities (1981).

The long-term contracts allegedly involved in *Commonwealth Edison*, permitting the producer to add additional taxes to the selling price, might make it easy to determine some minimum portion of the tax exported at the time it took effect, but as parties exercised rights under the contracts (*e.g.*, buyers exercising rights to reduce their purchases) and as the contracts expired, this simple calculation would become obsolete.

181 *See* Hellerstein, *Constitutional Constraints on State and Local Taxation of Energy-Resources*, 31 Nat. Tax. J. 245, 251 (1978).

182 The rule urged by the dissenters would have operated in this fashion. 453 U.S. 609, 651-53 (1981).

183 *See, e.g.*, the Court's rejection of such a rule for racial discrimination under the Equal Protection Clause, in Washington v. Davis, 426 U.S. 229, 248 (1976); *see also* Ely, *Legislative and Administrative Motivation in Constitutional Law*, 79 Yale L. J. 1205, 1254-61 (1970).

184 A perhaps startling example, in connection with the negative commerce clause, would be a local rent-control ordinance, which will have a differential adverse impact on nonresidents seeking housing in the locality, for it denies them the ability to bid against local residents for existing housing; by contrast, local resident tenants receive a quasi-property right in excess of what they bargained for in their leases. *See* Kitch, *Regulation and the American Common Market*, in Regulation, Federalism, and Interstate Commerce 31 (A.D. Tarlock ed. 1981).

The majority, besides rejecting the dissenters' approach, seemed implicitly to take the view that courts should find unconstitutional discrimination against interstate commerce only when it is express. This view of discrimination is not only too narrow, but deviates from the Court's normal view of the matter. There is a class of cases where the relation between the statute and the economic circumstances — and here we mean only rather obvious, easily ascertained economic circumstances — is extreme enough to justify a judicial finding that the legislature must have acted with an intent to discriminate against interstate commerce.

In some instances that relation makes the finding of discriminatory intent virtually inescapable. In 1971, for example, Ohio imposed a tax on the burning of coal, with the rates *falling* as the sulphur content of the coal *rose*. As the health hazards of coal-use rise in proportion to the sulphur content, the rate structure was suspect, at best. It turned out, moreover, that almost all Ohio coal was high-sulphur. According to one survey, only 4 percent of Ohio coal fell into the low-sulphur category that would be subject to rates of 30 cents a ton or more; the remaining 96 percent of reserves contained enough sulphur to enjoy a rate of only 15 cents per ton.[185] For practical purposes, then, low-sulphur coal (which the legislature had selected for especially heavy taxes) meant coal imported into Ohio.

In litigation over the tax Ohio offered no justification whatsoever for the peculiar rate structure. It defended the tax itself in terms of its use of the revenue to finance research into the better (and healthier) use of coal as an energy source; but, of course, that justification did not speak to the structure of the rates.

The courts — lower courts in this case — held the tax invalid. The opinions invoked the notion of discriminatory effect.[186] Since a true effect test is not really feasible, it would have been more candid and accurate to describe the cases as ones where (1) the relation between the statute and the circumstances, and (2) the lack of any nondiscriminatory explanation for the line drawn, compelled an inference of an intent to disfavour out-of-state suppliers of coal.

There is a closely related set of cases where the line drawn by the state does not compel an inference of discriminatory purpose, but almost does so. A recent example, drawn from the Court's commerce clause review of state regulation rather than taxes, is *Hunt v. Washington State Apple Advertising Commission.*[187] North Carolina had adopted a statute forbidding the sale within North Carolina of apples bearing any grade *other than* that of the United States Department of Agriculture. It had no grading system at all, while Washington, producer of 30 percent of the nation's apples, had a stringent inspection pro-

[185] *See* Mapco, Inc. v. Grunder, 12 Env't Rep. Cas. (BNA) 2025, 2026, 2029 (N.D. Ohio 1979).

[186] *See id.*; Dayton Power & Light Co. v. Lindley, 58 Ohio St. 2d 465, 391 N.E.2d 716 (1979).

[187] 432 U.S. 333 (1977).

gramme and a nationwide advertising programme. Washington's grading system complemented its advertising programme, since the state grade on the container alerted potential buyers to the provenance of the apples and perhaps implicitly suggested Washington's concern for high standards.

North Carolina defended the statute as intended to protect consumers from confusion or deception, but the Court found that the statute did little to fulfill that intention: (1) It was of little use against deception, since it allowed apples to be sold without any grading at all. (2) Since Washington's grades were equal or superior to their United States equivalents, they could not mislead a consumer to his detriment but only to his benefit. (3) A solution free of adverse impact on out-of-state apples was readily available: to require apple sellers to include the United States label.

The invalidation of state measures such as the Ohio coal-use tax and the North Carolina apple-labelling rule seem almost essential if a constitutional ban on *express* discrimination is to be meaningful. Unless the Court is ready to strike down legislation that is inescapably, or almost inescapably, referable to a discriminatory purpose, any rule against express discrimination will be too readily circumvented. Thus, such decisions function as a backstop to rules against express discrimination.

Can the Montana severance tax be distinguished from these cases? One way of formulating the question is to ask whether one can imagine a state enacting a 30 percent coal severance tax if it consumed all its own coal production. Surely the answer is affirmative. Strip-mining of coal entails some serious problems for a state: the environmental costs of the mining operations; the social, environmental and economic costs of the original boom; and the consequences that may befall the mining region when and if mining ceases to be profitable. Quite apart from these justifications, the tax is plausible as a legislative response to the inviting target of real or imagined economic rents.[188] European experience points in the same direction. The United Kingdom Government imposed a 20 percent severance tax on North Sea oil (in addition to a range of other taxes) at a time when the great bulk of production was for domestic use;[189] German Länder have imposed severance taxes — now at the level of 32 percent — on onshore oil production, all of which goes for domestic use.[190] All these make such a tax quite plausible within the framework of a single, economically isolated jurisdiction. It is virtually impossible to say the same of either the Ohio coal-use tax or the North Carolina apple-labelling rule. On such a record, invalidation of the Montana tax, merely because of the tax exportation element, would constitute a substantial judicial invasion of the

188 *See infra* Ch. III, pp. 116-24 for discussion of federal rent capture efforts, even in the face of a world trading situation such that the efforts tended to increase national dependence upon foreign suppliers.

189 *See infra* text accompanying notes 238-43.

190 "West German Länder to Raise Gas and Oil Royalties to 32 per cent," Financial Times, 12 Nov. 1981, at 3.

state's power to protect its environment and to guard against decline or the exhaustion of coal resources located within its borders.

Before leaving the subject of discriminatory purpose, we should distinguish (a) cases where such a purpose is inferred from the relation of the statute to the facts, from (b) cases where it is inferred from the legislative process itself — committee reports, statements of sponsors, and so forth. Many objections have been made to finding discrimination on the latter basis,[191] but there is one that seems to virtually dispose of the matter. If courts invalidate legislation solely on the basis of the legislative process, two anomalies would develop. First, other states, seeing the fate of the one whose statute was struck down, would be in a position to circumvent the decision. They could enact a similar statute, but, by keeping the record "clean", *i.e.*, free of allusions to discriminatory purpose, they could achieve what the first state could not. This would create an incongruous relation between the two states. Second, the first state might cure its problem by reenacting the statute on a clean record. But this, while curing the first anomaly, seems to dissolve the rule into a mere injunction to legislatures to act hypocritically. That is perhaps the last injunction legislatures need.

Neither of the main opinions in *Commonwealth Edison* adopted a clear position on proving discrimination through legislators' statements. The majority ignored plaintiffs' assertions on that issue altogether. The dissenters, while alluding to the alleged statements,[192] and making vague references to the risk of taxes "tailored" to reach interstate commerce,[193] gave no clue as to what role, if any, evidence from the legislative process should play.

Apart from *Commonwealth Edison,* the Court has occasionally [194] *mentioned* legislative assertions of discriminatory purpose, but we know of none where those assertions were the sole basis for a finding of discrimination, or transformed mere discriminatory effect into a case of constitutionally invalid discrimination. If the Court ever did make such evidence pivotal, the anomalies noted above would result.

Where does that leave *Complete Auto*'s prohibition against discrimination? Despite the majority's silence on the issue, we suspect that the Court would find state taxes discriminatory against interstate commerce on about the same basis that it finds state regulations discriminatory, as in *Hunt v. Washington Apple Advertising Commission.*[195] Thus, it would do so either if the tax were expressly discriminatory or if the relation between the tax and economic realities compelled, or very nearly compelled, the inference that the legislature in-

[191] *See especially* Ely, *supra* note 183, at 1212-17.

[192] 453 U.S. 609, 639-41 (1981).

[193] *Id.* at 649-50.

[194] *See, e.g.,* Hunt v. Washington State Apple Advertising Comm'n, 432 U.S. 333, 352 (1977); Village of Arlington Heights v. Metropolitan Housing Dev. Council, 429 U.S. 252 (1977) (racial discrimination context).

[195] 432 U.S. 333 (1977).

tended to discriminate against interstate commerce. But neither exportation of the tax, nor exportation combined with legislators claiming that the state could export the tax, is likely to violate the rule against discrimination.

(c) "Fair Relation" to Services Provided by the Taxing State

In the decision of the *Commonwealth Edison* majority, the requirement of *Complete Auto* that the tax be "fairly related to the services provided by the state" faded into thin air. The dissenters seemed to assign the test a contingent role: if the taxpayers could show that the tax was "exported",[196] then the tax could be saved either by proof (a) that it was roughly identical to taxes levied on "similar" industries, or (b) that the legislature could have found the tax "necessary to compensate the State for the particular costs imposed by the activity."[197] If one puts aside matters of proof, measurement, and the allocation of powers between state legislatures and the federal judiciary, the dissenters' rule would seem to make eminent sense. Either type of justification would tend to undercut suspicions that the prospect of exporting the tax had skewed the legislative process leading to its adoption. If the tax were at the same level as those applied to similar industries (where the tax was not exported), then one could infer that the legislature's selection of the level had nothing to do with the exportation of the tax. If the tax were needed to compensate the state for special costs, then it would be justified, like a pollution tax, as an effort to internalise external costs;[198] again its level would not seem attributable to distortions in the political process.

The actual hazards in the application of such a test, however, tend to undermine its theoretical appeal. Montana briefly listed the benefits that it provided the coal-mining industry and the costs that the industry inflicted on Montana, and claimed that they were unquantifiable.[199] A look at Montana's list reveals the sort of inquiry on which the dissenters would have launched the Court. First, to describe some of the benefits it provided the industry, Montana used the very words the Court itself had recently used to characterise the benefits that a state affords in exchange for taxes: "the benefits of a trained work force

[196] Exportation, in the view of the dissent, seemed equivalent to a prima facie case of discrimination.

[197] 453 U.S. at 651-52. The dissenters did not reveal on which party they would place the burdens of proof and persuasion. In *id.* at 652, the dissenters may have intended to suggest that even at this stage the tax would enjoy the presumptions that conventionally operate in favour of duly enacted legislation.

[198] Although a tax is not necessarily the optimal answer to the existence of an externality, *see* Coase, *The Problem of Social Cost*, 3 J. LAW & ECON. 1 (1960), the dissenters clearly regarded it as sufficient for these purposes. The work of Coase, however, suggests a difficulty with their proposed calculation in addition to those discussed in the text below. In identifying the "costs" imposed by the taxed activity, to what extent should the court include damages that would not have occurred but for the victims' failure to take cost-effective preventive measures?

[199] Brief of Appellees at 22-23, 26-27.

and the advantages of a civilized society."[200] It added a little precision by mentioning, and rhetorically asking how one could value, "the benefits of police and fire protection, the elborate network of roads and waterways, or the orderly system of laws that Montana provides."[201] Even if one sidestepped the problem of evaluating benefits by asking how coal-mining activity increased the *cost* of providing such benefits, it is clear that an answer would require arbitrary allocations.

Montana then pointed to a series of costs it said were specifically referable to coal-mining: (a) physical environmental costs, such as injuries to aesthetic values and to water quality; (b) social environmental costs — "increase in crime, alcoholism, and divorce"; and (c) the exhaustion of "a nonrenewable resource that constitutes the state's economic base."[202]

Not only is the quantification of environmental and social values inherently difficult, but it is integrally related to another highly intangible value of the federal system: the ability of different communities to reach diverse conclusions upon the right trade-off between such values as environmental quality and community stability, on the one hand, and pecuniary income and economic dynamism, on the other. The geographic and cultural diversity of the United States requires local autonomy on that trade-off; second-guessing by the federal courts would jeopardise both the autonomy and the resulting diversity.[203]

Any effective enforcement of a "fair relation" test would then involve the federal courts in measurement problems, arbitrary line-drawing, and invasion of state policy judgments very much akin to (and indeed overlapping) those entailed by a principle of invalidating taxes for their discriminatory effect. The Court's *de facto* abandonment of the test is, therefore, hardly surprising.

(d) Multiple Taxation of Interstate Commerce: The "Substantial Nexus" and "Fair Apportionment" Tests

A theme of the Court's supervision of state taxation has been that although interstate commerce may be made to "pay its way", it should "not be burdened with cumulative exactions which are not similarly laid upon local business."[204] If an enterprise has connections with more than one state, each may try to tax it on the basis of that connection; cumulatively the taxes may add up to more than those paid by a competitor active in only one state.

200 Japan Line, Ltd. v. County of Los Angeles, 441 U.S. 434, 445 (1979).

201 Brief of Appellees at 26.

202 *Id.* at 26-27. Montana's enumeration suggests some likelihood of double-counting: for example, the value of the police "benefits" that Montana provides the coal-mining industry must to a degree overlap the extra "costs" of crime attributable to the industry's existence and effects on the social fabric.

203 *See* Stewart, *Interstate Resource Conflicts: The Role of the Federal Courts,* 6 HARV. ENV. L. REV. 241 (1982).

204 Western Live Stock v. Bureau of Revenue, 303 U.S. 250, 258 (1938).

In the enforcement of this doctrine the Court has invalidated (1) taxes imposed by states whose relationship to the transaction was slight (or with which the taxing state lacked an adequate "nexus"),[205] and (2) taxes inadequately apportioned to take account of the contribution that the taxpayer's activities in other states made to the value taxed.[206]

The Court cannot, however, as a practical matter, fully achieve the goal of avoiding multiple taxation. So long as any two states tax differently, economic events straddling two jurisdictions are likely to be exposed to a higher tax burden than equivalent transactions entirely within one state. If, for example, State A relied exclusively on a property tax and State B exclusively on a sales tax, goods manufactured in A and sold in B would in effect be subject to two tax burdens, while goods manufactured and sold entirely in one or the other would be subject to only one. Thus the Court's goal could be fully achieved only if some central institution — the Court or Congress — imposed a single, unitary system of taxation upon all states.[207]

A surprising contrast between the Community and the United States is that the Community, despite its generally lower level of legal and economic integration, has moved more forcefully to solve this problem. It has harmonised the Member States' general indirect taxes, requiring that these conform to the model of a value-added tax, first set out in considerable detail in 1967[208] and made even more precise in 1977.[209] The scheme of the tax eliminates the possibility of double taxation.[210] It is imposed at each stage of the production and distribution cycle. Each taxable person invoices his customer with the value of the good plus the tax on that value, and accounts to the tax authorities at regular intervals for the tax he has thus collected on his outputs less the tax he has paid on his inputs. Exports, however, are "zero-rated", that is to say, the exporter passes on no tax to his foreign buyer, and can claim back from the tax authorities the tax he has paid on his inputs. The importer, conversely, is required to add to the value of the product when he sells it the appropriate sum for tax, which he will pay over to *his* tax authorities. The effect of this system,

205 *See, e.g.*, Norton Co. v. Department of Revenue, 340 U.S. 537 (1951); National Bellas Hass, Inc. v. Department of Revenue, 386 U.S. 753 (1967).

206 *See, e.g.*, J.D. Adams Mfg. Co. v. Storen, 304 U.S. 307, 311 (1938); Central Greyhound Lines, Inc. v. Mealey, 334 U.S. 653, 663-64 (1948); Evco v. Jones, 409 U.S. 91, 93 (1972).

207 A further difficulty with the goal of eradicating multiple taxation is that it lacks any secure starting point, in terms of what system, for a single jurisdiction, would constitute the baseline. States vary in the degree to which they slice the economic process into taxable incidents, and without some concept of a "normal" degree of slicing, one cannot safely say when multiple taxation begins.

208 Council Directive (EEC) 67/228 of 11 April 1967, 1967 J.O. 1303 (14 Apr. 1967), 1967 O.J. (spec. Eng. ed.) 16.

209 Council Directive (EEC) 77/388, O.J. L 145, 13 June 1977, p.1, art. 37, which repealed the 1967 Directive.

210 *See*, for full discussion, A.J. Easson, Tax Law and Policy in the EEC (1980).

which is an example of what is generally termed the "destination" system of indirect taxation, is that on an exported product the whole tax revenue is collected by the importing state; the exporting state gets nothing, because it remits to the exporter the whole of the tax burden being carried by the product up to that stage; and the product is thus taxed once only.

This system has worked well, though it depends on the willingness of Member States to view their tax losses on exports as being compensated by their tax gains on imports, or by other economic advantages, a condition which may not obtain in the United States. Moreover, it requires the maintenance of tax frontiers, and it remains the objective of the Community to abolish these, which implies a switch to an "origin" system of taxation with uniformity of rates. Finally, we should note that the system does not preclude the operation of specialised indirect taxes, like excises, severance taxes, and parafiscal levies. These, as we shall see,[211] continue to cause problems to the Community.

Efforts by the United States Supreme Court to achieve such uniformity would surely be viewed as intolerable instrusions upon the powers of the states. Accordingly, it is not surprising that the Court's efforts in this direction have been halting. Gross receipts taxes — of which a mineral serverance tax is a variety — raise the issue acutely. Gross receipts represent an accumulation of values ranging from initial production through to final sale. Where the components of the process are in reality spread over more than one state, the principle of avoiding multistate taxation would seem to require apportionment. But the Court has by no means always imposed that requirement. It has, for example, allowed a state into which new cars were delivered to impose an unapportioned tax on the gross receipts from sale of the cars,[212] although surely the taxpayer's activities in the state of manufacture made a considerable contribution to the value taxed. And it has allowed a similar tax on gross receipts in the state of manufacture.[213]

Such gaps as these effectively permit the states to escape the apportionment requirement merely by identifying some in-state slice of the economic process, and taxing the gross income received at that stage. Indeed, until the decision in *Commonwealth Edison*, commentators noting these gaps used to point to

[211] *See infra* pp. 82-84.

[212] General Motors Corp. v. Washington, 377 U.S. 436 (1964); for a critical comment on the Court's attitude toward the Washington tax, *see* Hellerstein, *State Taxation of Interstate Business and the Supreme Court, 1974 Term:* Standard Pressed Steel *and* Colonial Pipeline, 62 VA. L. REV. 149, 168-76 (1976).

[213] Although the state of manufacture may not impose an unapportioned tax directly on gross receipts, J.D. Adams Mfg. Co. v. Storen, 304 U.S. 307 (1938), it may impose a licence tax on manufacturing and measure the tax by the resulting gross receipts (notwithstanding interstate shipment of some of the goods), American Mfg. Co. v. St. Louis, 250 U.S. 459 (1919).

Heisler as an example of ways a state could artificially escape the requirement.[214]

The *Commonwealth Edison* decision, though broadly condemning the "mechanical" character of the *Heisler* approach, and purporting to commit the Court to analyse mineral extraction taxes in terms of their "practical effect",[215] seemed blind to the multiple taxation hazards of a severance tax. Although plaintiffs did not challenge the want of apportionment, the Court affirmatively claimed that none was needed.[216] The attitude appears to be part of a more general disposition of the Court to relax its concern over multiple taxation — perhaps out of a recognition that in this area its reach has exceeded its grasp.[217]

(e) Conclusion

Judicial intervention against state taxation of energy resources, then, seems largely limited to striking down taxes that are (a) explicitly discriminatory (or whose structure compels an inference of discriminatory purpose) and (b) egregious intrusions upon its goal of limiting multiple taxation.

2. In the European Community

As in relation to other devices which could limit free trade, the Community Treaties spell out the relevant prohibitions regulating taxation in greater detail than does the US Constitution. Again, however, explicitness in Treaty formulation has not disposed of all questions regarding the tax regime.

Taking the simplest case first, one which, indeed, is not even present now within the United States system, customs duties on exports are forbidden within all three Communities. Under article 16 of the EEC Treaty, customs duties on exports, and charges having equivalent effect, were to be abolished in the EEC "by the end of the first stage", that is, by 31 December 1961.[218] In Euratom, they were to be abolished within a year from the coming into effect of the Treaty.[219] In the ECSC, export duties, and charges having equivalent effect, were "recognised as incompatible with the common market" and were to "be abolished and prohibited within the Community"[220] from the date of the establishment of the common market in the relevant product which, for coal, was 10 February 1953. When perfected by time, these were intended as

214 *See* Powell, *More Ado About Gross Receipts Taxes,* 60 HARV. L. REV. 501, 534-35, 721-22 (1947); Hellerstein, *supra* note 212, at 174.

215 453 U.S. 609, 614-16 (1981).

216 *Id.* at 617.

217 *See, e.g.,* General Motors Corp. v. Washington, 377 U.S. 436 (1964); Moorman Mfg. Co. v. Bair, 437 U.S. 267 (1978) (allowing states a wide range of freedom in selecting methods of apportionment, despite the risks of multiple taxation implicit in such freedom).

218 EEC Treaty art. 16.

219 Euratom Treaty art. 93.

220 ECSC Treaty art. 4(a).

absolute prohibitions; in particular it should be noted that whereas article 36 of the EEC Treaty allowed certain derogations from the Treaty prohibition of quantitative restrictions on trade, on such grounds as public morality, public policy and public security, this derogation did not apply in relation to customs duties, whether on exports or imports.[221]

It will be noted that in each of the Treaties, the prohibition on customs duties extends also to "charges of equivalent effect". This phrase appears in all three Treaties,[222] but only in the EEC Treaty has it given rise to litigation in which the Court of Justice has been able to offer an authoritative interpretation. In the case of *Commission v. Italy* in 1969, the Court offered a general definition of such charges, applicable both to those on imports and those on exports, in the following terms:

any pecuniary charge, however small and whatever its designation and mode of application, which is imposed unilaterally on domestic or foreign goods by reason of the fact that they cross a frontier, and which is not a customs duty in the strict sense ... even if it is not imposed for the benefit of the State, is not discriminatory or protective in effect and if the product on which the charge is imposed is not in competition with any domestic product.[223]

Such a broad definition might have made the application of the Treaty provisions unproblematical, were it not for the fact that the EEC Treaty, unlike the other Treaties, creates a separate regime for internal taxation, which is governed by articles 96-99. Unlike article 16, which forms part of the title "free movement of goods", these articles, which figure in the common policy part of the Treaty, are directed not at the burdening of inter-state trade as such but rather at discrimination in internal tax systems. Articles 95 and 96, the basic provisions, are designed on the assumption that Member States will apply the destination principle in relation to the internal taxation of imports and exports: that is to say, that they will tax imports and, on the assumption that other Member States will do the same, will remit internal taxation on their exports. The dangers foreseen were that Member States, in the operation of this system, would overtax imports by imposing rates of tax higher than on competing domestic products, and subsidise exports by making tax repayments in excess of the internal taxation imposed on them. Article 95 forbids the former practice,[224] article 96 the latter. Neither article, however, explictly forbids the Member State to penalise its exports by denying or reducing repayments of in-

[221] *See* Case 7/68, Commission v. Italy, [1968] E.C.R. 423. Article 36 provides that "the provisions of Articles 30 to 34 shall not preclude prohibitions or restrictions on imports, exports or goods in transit justified on grounds of public morality, public policy or public security...."

[222] EEC Treaty art. 16; ECSC Treaty art. 4(a); Euratom Treaty art. 93.

[223] Case 24/68, [1969] E.C.R. 193, 201. The last words of the quotation, of course, are apt only in relation to imports.

[224] For an example of discriminatory internal taxation involving a natural resource (timber), *see* Case 77/69, Commission v. Belgium, [1970] E.C.R. 237.

ternal taxation (which, in the context of trade between states applying the destination principle, may result in double taxation) or by imposing taxes on certain goods, within the framework of a system of internal taxation, only when intended for export as apposed to domestic consumption.

In regard to charges which burden exports, therefore, the regimes of articles 16 and 95-99 appear to point in opposite directions: article 16 forbids them, articles 95-99, if they form part of a regime of internal taxation, permit them. Two cases presented to the Court in 1974 required it to resolve this conundrum. The relationship between articles 16 and 96 was first treated in *Demag AG v. Finanzamt Duisburg-Süd.*[225] In line with the assumptions of article 96, the turnover tax regime in Germany provided for the repayment, at the border, of tax estimated to have been paid on German exports. In 1968, however, in order to redress an excessive balance of payments surplus, a temporary amendment was made to the turnover tax law, granting tax reliefs on imports and imposing on exports enjoying turnover tax rebates a temporary tax levied at 2 or 4 percent (according to the level of rebate). Confronted with the question of the legality of this type of tax under the EEC Treaty, the European Court first held that a charge could not at the same time be a charge of equivalent effect under articles 12 and 16, and an element of the Member State's regime of internal taxation regulated by articles 95-99 of the Treaty. It further held that a tax law of the type in question, which could be seen as cancelling a rebate ordinarily enjoyed under the turnover tax law, should be held to be so closely associated with the turnover tax regime as to fall within articles 95-99 and, therefore, outside article 16. Since article 96 did not explicitly forbid the non-remission of export taxes, no breach of the Treaty had occurred. After *Demag*, therefore, the factor determining the legitimacy of a tax effectively burdening exports appeared clearly to be its allocation to the regime either of article 16 or of article 96.

Only a few weeks later, however, the Court's decision in *P.J. Van der Hulst's Zonen v. Produktschap voor Siergewassen*[226] suggested that it was taking a broad view of what constituted a charge of equivalent effect under article 16, without worrying too much about whether the charge would also be qualified as internal taxation under articles 95-99. In this case the plaintiff exporters attacked two levies imposed on sales of Dutch bulbs, within the framework of a comprehensive scheme of regulation of bulb marketing. The first levy, which financed the disposal of bulb surpluses, was paid by all sellers of Dutch bulbs, and presumably passed on by them, but could effectively be reclaimed from the marketing authority by Dutch purchasers, though not by foreign purchasers. The second was paid at a lower rate on domestic than on foreign sales, and served to fund the promotion of the Dutch bulb trade. Here the Court held that an internal levy falling more heavily on exports than on internal sales, or intended to finance activities likely to make internal marketing more

[225] Case 27/74, [1974] E.C.R. 1037.
[226] Case 51/74, [1975] E.C.R. 79.

profitable than exporting, or otherwise to give preferential treatment to the product destined for the domestic as opposed to the export market, might have an effect equivalent to a customs duty on exports, and, therefore, be in breach of article 16. Articles 95-99 were not explicitly argued by the parties or the Advocate-General, but the Court, in holding that the levies also contravened the relevant EEC agricultural regulation, thought them incompatible, if only by analogy, with prohibitions on discrimination including that contained in article 95, because exported goods were subject to a heavier charge than that placed on the internal market, or because the revenue from the charge was intended to place national products at an advantage.

Hulst showed the Court to be prepared to take a broad view of the application of article 16. The discriminatory nature of the charge was here clear, but it is worth noting, particularly in this comparative context, that the Court was prepared to look at apparently nondiscriminatory levies, at least if they produced a discriminatory effect by being *applied* for the exclusive benefit of the national product. In pursuing this line the Court was following its earlier jurisprudence in relation to charges on imports, on which it had held[227] that apparently nondiscriminatory internal charges on imports could constitute charges of equivalent effect to import duties under article 9 if their proceeds were applied for the benefit of the domestic product. Clearly, a straight repayment of the charge to the domestic producer, as in *Hulst*, falls foul of this principle, but there is room for argument as to how specific the link between the proceeds of the charge and the provision of benefits to the domestic producer needs to be. Earmarking of the proceeds for some purpose of benefit to the domestic industry would seem essential; obviously every internal tax, by contributing to public revenue, benefits those resident in the state largely to the exclusion of those not there resident, but this can hardly be the basis for attacking a charge falling equally on production for home consumption and for export, even, it would appear, if it can be shown that the burden of the tax is effectively exported.[228]

The results of the *Demag* and *Hulst* cases now need to be seen in the light of the Court's further, and remarkable, decision in *Statens Kontrol med Ædle Metaller v. Preben Larsen*, in 1978.[229] This case involved a Danish tax on precious metals, which was applicable to the total consumption of enterprises working such metals, whether the articles made were sold in Denmark or exported. Assessment to tax for articles sold within Denmark was based on the quantities marked with the hallmark of the enterprise as required by law, un-

[227] Case 77/72, Carmine Capolongo v. Azienda Agricola Maya, [1973] E.C.R. 611. The same issue was discussed by the Court in a second and third case on the ENCC levy, Case 94/74, Industria Gomma Articoli Vari, IGAV v. Ente Nazionale per la cellulosa e per la carta, ENCC, [1975] E.C.R. 699 and Case 74/76 Iannelli & Volpi SpA v. Ditta Paolo Meroni, [1977] E.C.R. 557.

[228] *See infra* text accompanying note 234.

[229] Case 142/77, [1978] E.C.R. 1543.

less they were sold to a buyer in Denmark with his own hallmark, in which case they would be included in his consumption for tax purposes. Even though exported articles were not subject to hallmarking, they still attracted the tax. Goldsmiths working articles for export attacked the tax as contrary both to articles 16 and 95.

Though the original purpose of the tax was to finance assay and hallmarking, the Court refused to see in this a benefit which would accrue to the domestic market only, an approach which would have led to the application of the discrimination tests for article 16 applied in *Hulst*. Instead it simply held that this was a regime of internal taxation, to which article 16 could not apply, because the duties were not imposed "specifically" on exported products.[230] That strict interpretation of article 16, one would have thought, closed the case. But no: noting the Treaty's lack of symmetry in the treatment of internal taxes on imports and exports, the Court went on to hold that it followed from a comparison of articles 95 and 96

> that the aim of the Treaty in this field is to guarantee generally the neutrality of systems of internal taxation with regard to intra-Community trade whenever an economic transaction going beyond the frontiers of a Member State at the same time constitutes the chargeable event giving rise to a fiscal charge within the context of such a system.
>
> It therefore seems necessary to interpret Article 95 as meaning that the rule against discrimination which forms the basis of that provision also applies when the export of a product constitutes, within the context of a system of internal taxation, the chargeable event giving rise to a fiscal charge.[231]

As a result of this remarkable piece of teleological interpretation the identification of a charge as within or outside a system of internal taxation, which seemed so vital in *Demag*, loses much of its significance. Now it seems that if a tax is explicitly discriminatory against exports, it may be caught under article 95 and, if *Hulst* retains any authority (and it should be noted that the Court cited it in *Ædle Metaller*[232]) under article 16 as well. If, on the other hand, the tax burdens both exports and production for internal use, it may still be caught under article 95, if discriminatory in effect. It should be noted that although the tax in question here was based on consumption of precious metals, and was not a tax imposed "specifically" on exports, it was nonetheless apparently regarded by the Court as a tax under whose regime "the export of a product constitutes ... the chargeable event", in that the Court went on to consider the question of discrimination. In so doing, however, it looked only at the question whether the exported product was treated by the tax regime in the same way as the product for internal use; it refused to consider the double taxation effect arising from the non-remission of the tax on export in situations where other states imposed hallmarking charges on the goods in question. Non-remission was, in the Court's view, a problem calling for legislative, not judicial,

[230] *Id.* at 1557, para. 16.
[231] *Id.* at 1558, paras. 23-24.
[232] *Id.* at 1558, para. 27.

solution.[233] Discriminatory effect, therefore, must be judged having regard only to the system within the taxing State.

Despite the boldness of the *Ædle Metaller* decision, it does not seem that the case-law of the Court of Justice goes much further than has that of the Supreme Court to eliminate the exportation of tax burdens between the states and the possibility of multiple taxation on certain economic activities. This is because the question whether the tax burden falls more heavily on export than on domestic sales is answered without having regard to the subsequent tax burdens on the exported good and without regard to the proportions of the good that are consumed at home and exported abroad. The question of exportation of the tax burden has not been explicitly raised in Europe as it was in the *Commonwealth Edison* case, but argument by analogy from the Court's jurisprudence on taxes on imports, under article 95, strongly suggests that even if there were *no* domestic consumption of a good produced within a Member State, it could still be the subject of a non-remitted export tax provided that that tax formed part of a general system of internal dues applied systematically to categories of products irrespective of the destination of the products.[234]

In practice this latter condition is unlikely to obtain, because, as we have seen, indirect taxes of general application have been harmonised within the European Community by the adoption of a common system of value-added taxation (VAT) which requires the remission of VAT on exports.[235] Member States remain free, however, to use other indirect taxes, and petroleum products in fact bear quite high (and varying) levels of excise duty in all countries of the Community.[236] If such an excise duty were to be imposed on export sales of a product without domestic consumption (or perhaps with very low domestic consumption) and not remitted, it might still be held to fall within a system of internal dues so as to avoid the obvious imputation of discrimination. The hypothesis is, in any event, fanciful: we are not aware of any product in the energy field which is wholly or almost wholly exported in this way. As with turnover taxes, the EEC is attempting to solve the problem of varying levels and systems of excise duties through harmonisation, though without notable success so far.[237] Perhaps such harmonisation would be the most sensible legislative response to the fears of tax exportation raised in the United States by the *Commonwealth Edison* decision.

233 *Id.* at 1559-60, paras. 33-36. The Treaty mechanism for this is to be found in the tax harmonisation provisions of art. 99.

234 This was the sense of the Court's ruling in relation to imports in Case 90/79, Commission v. French Republic (Reprographic Machines), [1981] E.C.R. 283, 301-02, paras. 14-15.

235 *Supra* pp. 75-76.

236 EC Commission, Taxation of Petroleum Products, Doc. COM(81) 511/final, Annex pp. A.2 & 12, Table 1 (11 Sept. 1981).

237 EC Commission, Proposal for a Council Directive on the Harmonization of Excise Duties on Mineral Oils, O.J. C 92, 31 Oct. 1973, p. 36, *especially* art. 9(1), providing for exemption from duty for exported products.

While the EEC Treaty thus leaves some leeway for tax exportation by resource-rich Member States, there is no evidence that such exportation or discrimination has occurred within this field. Specially designed resource taxation is most unusual; mining operations are subject to the general regimes of income or corporate taxation, which have neither the object nor the effect of discriminating between production destined for export and for domestic consumption.[238] Within the range of resources subject to the EEC Treaty, the most significant special measures of resource taxation are the United Kingdom Petroleum Revenue Tax and Supplementary Petroleum Duty. The former is in the nature of a specialised profits tax, being applicable only to profits from oil accruing to persons participating in oil fields in the United Kingdom, its territorial sea and continental shelf.[239] Profits are assessed on a field-by-field basis according to a complex formula;[240] in this assessment, the destination of the oil sold by the participants in oil fields is of no significance.[241] The assessment is simply based on the proceeds of sale of oil sold crude (the sale price if sold at arm's length, the market value if not) by each participant in a given period. The Supplementary Petroleum Duty, which applied only for the two years 1981-82,[242] was additional to Petroleum Revenue Tax and was charged at 20 percent on "gross profit" as defined for Petroleum Revenue Tax purposes.[243] In fact "gross profit" so defined is broadly equivalent to the market value of the petroleum produced, so that Supplementary Petroleum Duty was, in effect, a severance tax similar to that considered by the Supreme Court in *Commonwealth Edison*. Like Petroleum Revenue Tax, however, it is assessed by reference not to specific sales transactions, but to oil income of each participator in an oil field over a period. This system excludes the possibility of an expressly discriminatory export regime, and has no "obviously" discriminatory effects of the type which led to the Court's finding in *Hulst*; it, therefore, cannot fall foul of the prohibition of article 16 or meet the test of *Ædle Metaller*. Moreover, neither of these United Kingdom taxes involves the improper exportation of the tax burden to consumers in other

[238] For a general survey, *see* TAXATION OF THE EXTRACTIVE INDUSTRIES (International Fiscal Association ed., vol. 63a Studies on International Fiscal Law/Cahiers de droit fiscal international, 1978) which includes reports on West Germany, Denmark, France, Italy, Netherlands and the United Kingdom.

[239] For details, *see* R.F. HAYLLAR & R. ROUSE, UK TAXATION OF OFFSHORE OIL AND GAS (2d ed., looseleaf, 1981); UNITED KINGDOM OIL AND GAS LAW, *supra* note 56, at paras. 1-1184.

[240] *See* Oil Taxation Act 1975, ch. 22, § 2 and Sch. 3.

[241] A minor exception is to be found in the Oil Taxation Act 1975, ch. 22, § 10(i)(a), but since it relates only to gas sold or to be sold under contracts concluded before the coming into force of the Act, it should have no effect on future behaviour.

[242] It was imposed by Finance Act 1981, ch. 35, § 122 and removed by Finance Act 1982, ch. 39, §§ 139-142, which instituted in its place a system of advance payment of Petroleun Revenue Tax.

[243] *See* Oil Taxation Act 1975, ch. 22, § 2(4) & (5).

Member States. North Sea crude, which for the most part has light, low-sulphur characteristics, competes for West European markets with a variety of non-EEC crudes with these characteristics, from Libya, Algeria, the Middle East, Mexico, Norway and so on. Term prices are made and maintained by coordinated action and conscious parallelism among producers; spot price variations respond to competitive pressures in the market. In neither case do United Kingdom producers, including BNOC with its large market share, enjoy an independence in pricing sufficient to enable them to pass any significant part of tax increases on to their foreign (or domestic) customers.

The relationship between internal taxation and charges having equivalent effect to export duties presented itself somewhat differently within the ECSC. That Treaty contains no provisions comparable to articles 95-99 of the EEC Treaty. Thus the ECSC Treaty lays down no restrictions, and confers no powers on the High Authority, in relation to the general tax regimes of the Member States. Indeed, at the time the Treaty came into force there was no agreement between the Member States as to the basic principles of indirect taxation, the Germans (incidentally the largest and lowest-cost producers and exporters of coal within the Community) favouring the origin principle, the rest the destination principle. The eventual agreement of the Member States, taken outside the framework of the Treaty, to adopt the destination principle[244] could not, however, have prevented the German authorities from pursuing an independent tax policy, involving, for example, the general limitation of tax refunds on exports. Limitations, or additional taxes, falling specifically on coal production or exports might, however, be suspect either under article 4(a) or under article 4(c), which forbids "special charges imposed by States, in any form whatsoever". This latter provision was applied by the High Authority in the early days of the ECSC in a situation a little like that in *Hulst*: the High Authority demanded and secured the discontinuance of the requirement placed on the German coal industry to supply coal at cheap rates to users such as German railways and power stations.[245] Such a requirement could have an effect similar to that of a levy on export and other "non-privileged" sales, for the benefit of the privileged consumers.

In fact these possibilities of export taxation have been of little significance within the ECSC, notwithstanding the leading position of the German coal industry within the Community between 1951 and United Kingdom entry in 1972.[246] This is because the High Authority had other instruments available to

[244] This agreement followed the submission of a report by a committee of experts (the *Tinbergen Committee*) commissioned by the High Authority; *see* HIGH AUTHORITY, REPORT ON THE PROBLEMS POSED BY THE DIFFERENT TAX SYSTEMS WITHIN THE COMMON MARKET (Brussels, 1953).

[245] Decision 25/53, 1953 J.O. C.E.C.A. 83; Decision 17/54, 1954 J.O. C.E.C.A. 266.

[246] Germany's share of Community coal production was about 63% in 1960, and had climbed to more than 73% immediately prior to the entry of the UK (figures calculated from OECD/IEA, ENERGY BALANCES OF OECD COUNTRIES 1960/1974 (1977)).

prevent any abuse by Germany of its market strength. Principal among these was direct control of maximum and minimum prices under article 61: a general control of maximum prices for coal was established in March 1953[247] and maintained at least in relation to the mines of the Ruhr, until 1956. In addition, the Convention on Transitional Provisions makes provision for equalisation schemes, both general and specific to particular States, designed to reduce the impact of differences in the costs of production in different areas of the ECSC and thereby to allow time for the running down of excessively costly production capacity.[248]

III. Corrections of Market Failures (or Offsets to Inadequate Specification of Property Rights)

A. In the United States

Inadequate definition of property rights causes misallocation of resources. Where the inadequacy cannot be cured directly, *i.e.*, where some feature of the resource makes the establishment of secure, definite and transferable property rights impossible or very costly, some government is likely to intervene and establish regulatory controls.

The intervention, naturally, provides an opportunity — the thin end of the wedge — for the accomplishment of other purposes. And those other purposes may come to dominate the intervention. Nevertheless, the problems in the specification of property rights tend to explain at least the form of the intervention.

1. "Common Pool" Problems

(a) Oil and Gas: Well-Spacing, Allowables and Market-Demand Prorationing

Under the Rule of Capture, the owner of land containing part of a pool of oil or gas is entitled to extract oil or gas from a well within that land, even though the mineral that he extracts originates in another's land. In the case of oil and gas in the ground, the obstacle to the adequate definition of property rights has been the fugacious character of the resource — withdrawal in one place

[247] Decision 6/53, 1953 J.O. C.E.C.A. 63. This Decision, it may be noted, permitted the inclusion of turnover taxes in export prices (art. 3), though a later Decision (Decision 30/53, 1953 J.O. C.E.C.A. 109, 1952-1958 O.J. (spec. Eng. ed.) 9, art. 5) made it clear that such inclusion was not permissible in any case where the tax was remitted or not payable.

[248] ECSC Treaty arts. 25-28.

causes other oil and gas to move about — coupled with inability to measure these movements precisely. Thus each owner's interest in the oil or gas in his land is defeasible by the action of his fellow owners in the pool. This gap in the owners' property rights generates various forms of wasteful behaviour. Two are critical for our purposes: the drilling of an excessive number of wells and extraction at too rapid a rate.

(i) *Excessive number of wells.* A well may extract oil or gas that others' wells would have produced anyway. A single owner of the entire field, considering investment in a new well, would clearly deduct production expected to be lost at any other well. He would proceed with a new well only if the expected increment in the *net* value of production exceeded the well's expected cost.[249] Under the Rule of Capture a mineral owner estimating production from a prospective well would not subtract losses in production at others' wells. Owners of interests in the common pool of minerals, therefore, are likely to drill wells whose costs exceed the net increment that they generate in the value of production from the pool.

(ii) *Too rapid a rate of extraction.* Under the Rule of Capture owners in a field tend to disregard a key cost of current extraction: the value lost as a result of foreclosing the possibility of extraction later.[250] An owner of property rights in oil or gas who anticipates rising prices, and concludes that the present value of the mineral would be greater if he started extraction five or twenty years in the future, cannot defer drilling on the basis of that calculation. What he does not extract today he may never be able to extract.

A further cost of rapid extraction is damage to the producing capacity of the reservoir. It will, for example, cause channnels to form between the water table and the well bore, rendering some oil unrecoverable as a practical matter. Under the Rule of Capture an owner enjoys all the benefits of rapid extraction but will suffer only a portion of reservoir damage; the result is of course excessive speed.

One cure for the problem is "unitisation", in effect a reshuffling of property rights so that decisions as to operation of the pool are made on a unified basis, taking into account all costs and benefits to owners in the pool. Under unitisation, each owner's return is no longer contingent upon the amount of oil or

[249] An "expected value" is in essence an average of the possible values that the factor may take, each weighted by the probability of its occurrence. Thus if a well has a 9:10 chance of being a dry hole and a 1:10 chance of producing oil with a discounted present value of $10 million the expected value of its production is $1 million.

Because of discounting to present value, additional wells that do not increase the aggregate *quantity* of production from a field may increase its *value*. For example, if one assumed a 5% discount rate, a constant price of oil, and no general price inflation, the present value of producing $500,000 worth a year for 20 years is $6.25 million, while the present value of producing $1 million a year for 10 years (twice as much annual production for half as long) is $7.72 million.

[250] *See* discussion *supra* pp. 31-32.

gas produced on his land, but rather on his pro rata share of the underlying mineral[251] and the extent to which he participates in the financial risks of drilling.

Unitisation can result simply from the agreement of all owners of interests in the pool. But a state may facilitate unitisation by providing that owners of a certain percentage of the interests can adopt unitisation over the resistance of the minority. The provision of such a device, called "compulsory unitisation", reduces the likelihood that individual owners will prevent unitisation by holding out for disproportionate shares of the resulting gains.

The second type of cure is direct regulation of the extraction process. Its most important features, for our purposes, are well-spacing requirements and "allowables". The former prohibit drilling of wells at greater than a specified density (say, one well per 40 acres); the latter constitute ceilings on the amount of production from each well (or spacing area) within a specified time-period.

As unitisation centralises decision-making in a body that must consider effects on all owners' interests (*i.e.*, "internalises the externalities"), its occurrence would justify dispensing with well-spacing and allowables regulation as to a unitised field.[252] In fact states do not provide for such dispensation. Because of that, and because many pools are not unitised, the programme of well-spacing and allowables regulation applies to production from all pools (in all the significant oil and gas producing states).[253]

Well-spacing and allowables regulation might aim at achieving, for each pool, the optimal rate of production — *i.e.*, the rate that a single owner of the pool aiming to maximise the net value of the resource would seek. For various reasons — the need for administrative simplicity, the high cost of data individualised for each pool, and perhaps reluctance to articulate the economically relevant data (such as discount rates, prices for oil and gas and for the inputs to production, and expected price trends) — state legislatures and administrators have not aimed at such a target. It appears that the closest that regulators may have gone towards aiming explicitly at optimal production rates is to aim to maximise *physical* recovery from the pool *subject* to a constraint of not letting the extractor's rate of return fall below some acceptable level.[254] Nonethe-

[251] Determining pro rata shares is not easy. Approaches vary from simple use of acreage to complex formulae that seek to take account of many factors bearing not only upon the quantity of mineral in each owner's tract but also on its contribution to ultimate production, which may be affected by other features. Obviously there is a trade-off between simplicity and exactness of correlation with each owner's contribution.

[252] *See* S.L. McDonald, *supra* note 15, at 201-09, 232-33, 244.

[253] *See id.* at 150-51. A few states do not control the rate of flow, but they are not ones whose production is important. *Id.*

[254] *See* S.L. McDonald, The Leasing of Federal Land for Fossil Fuels Production 128-30 (1979). *See also* S.L. McDonald, *supra* note 15, at 150-96, especially his discussion (at 184) of the idea of a ceiling rate of production formulated as the rate that cannot be exceeded without incurring "some (technically) avoidable loss of ultimate recovery."

less, it is probably fair to describe the regulatory system described above as a genuine, if crude and arbitrary, endeavour to offset the distortions caused by imperfect specification of property rights in oil and gas.

Five states, accounting for about three-fourths of liquid hydrocarbon production in 1970, however, have in the past superimposed an additional type of restriction — "market demand prorationing" or "MDP".[255] This originated in Oklahoma in 1927[256] and effectively ended in 1972.[257] It involves restricting production at each well to some fraction of its nominal allowable, in the supposed interest of equating supply with demand. Since there is always some price — the market-clearing price — that will achieve that equation, MDP in fact represented indirect determination of price by the five states. Accordingly, there is ample ground for suspicion that they used MDP to secure monopoly profits for producers, at the expense of consumers.

In any event, in 1928 the Supreme Court sustained Oklahoma's regulatory programme — both its probably innocent well-spacing and allowables regulation and its not-so-innocent MDP — against a negative commerce clause attack. In *Champlin Refining Co. v. Corporation Commission of Oklahoma*,[258] the Court dispatched the negative commerce clause claim in one short paragraph, asserting that the orders applied "only to production and not to sales or transportation of crude oil". "Such production is essentially a mining operation and therefore is not part of interstate commerce even though the product obtained is intended to be and is in fact immediately shipped in such commerce."[259]

The opinion left ambiguous whether the Court would accept such production controls where it recognised that they were ancillary to a programme of sustaining prices. Just ten years before, in *Lemke v. Farmers Gain Co.*,[260] the Court had struck down North Dakota's minimum price scheme for grain. Could a state do by production controls what it was forbidden to do by price floors? The *Champlin* opinion underscored the uncertainty by being resolutely obtuse on the relation between price and quantity. Although the challenger had forcefully argued that the quantity demanded was contingent on price, so that the Oklahoma regulatory commission's findings on the quantity demanded necessarily assumed a price (and so that its setting production to match the quantity necessarily *set* a price),[261] the Court (elsewhere in the opinion) cited with approval findings below that "none of the proration orders here involved were made for the purpose of fixing prices."[262]

255 The states are Texas, Louisiana, New Mexico, Oklahoma and Kansas. *See* S.L. McDonald, *supra* note 15, at 151.

256 *See id.* at 36-37.

257 *See* J.P. Kalt, The Economics and Politics of Oil Price Regulation 8, 297 (1981).

258 286 U.S. 210 (1928).

259 *Id.* at 235.

260 258 U.S. 50 (1922).

261 286 U.S. at 215-16.

262 *Id.* at 234.

A later decision outside the energy field suggests that the Court would probably uphold such a programme even where it fully recognised the relation between fixing production quotas and fixing price. In *Parker v. Brown*,[263] it upheld a California raisin-marketing scheme aimed at reducing the number of raisins reaching the market. The price-raising goals of the scheme were candidly acknowledged, albeit clothed in the usual glossy raiment of such programmes: an effort to assure "parity" prices, or prices that "would bring a fair return to the producers", or prices that would prevent "demoralization of the industry".[264] The Court's opinion made no effort to penetrate these generalities. Although California exported 95 percent of its raisin crops,[265] the Court found "the matter one which may appropriately be regulated in the interest of the safety, health and well-being of local communities, and which, because of its local character, and the practical difficulties involved, may never be dealt with by Congress."[266] Thus the Court's modern-day tolerance of production ceilings is probably not based on blindness to their relation to price floors, but upon an abandonment of *Lemke v. Farmers Grain Co.* and, perhaps, a generally less intrusive attitude.[267]

The legal status of state well-spacing, allowables, and market demand prorationing is further complicated by federal adoption of the "Connally Hot-Oil Act"[268] and authorisation of an Interstate Compact to conserve oil and gas.[269] The former makes it a federal offence to engage in the interstate transportation of oil that is "contraband" under state law, and was enacted in 1935 with full understanding of the existence of market demand prorationing. The Compact formalises interstate cooperation relating to the conservation of oil and gas.[270] The Connally Act, at least, could give rise to an argument that Congress had exercised its commerce clause power to approve market demand prorationing by the states, thus nullifying any claim under the negative commerce clause.

[263] 317 U.S. 341 (1943).
[264] *Id.* at 364, 367.
[265] *Id.* at 359.
[266] *Id.* at 362-63.
[267] Kommers & Waelbroeck, *supra* note 3, at 176-79, 190-96.
[268] 15 U.S.C. paras. 715-715d.
[269] Congress most recently consented to the pact in 1976, P.L. No. 94-493, 90 Stat. 2365. The consenting resolution states that the consent is "to an extension and renewal from September 1, 1974, to December 31, 1978", but the pact itself states that it "shall continue in effect until Congress withdraws its consent." Whatever the meaning of these provisions, the Compact may be valid for the period after 1978 even if one assumes that Congressional consent expired on 31 December 1978, for certain compacts are understood not to require such consent. *See* United States Steel Corp. v. Multistate Tax Commission, 434 U.S. 452 (1978).
[270] Although it states in article V that it does not authorise limitations on the production of oil or gas for the purpose of stabilising or fixing the price, neither this clause nor the Compact as a whole represents a repudiation of market-demand pro-rationing, especially as the proponents of such pro-rationing deny that it is in any way aimed at price-fixing.

(b) Water: The Non-Ownership of Unappropriated Waters

Under the system of prior appropriation, no one can acquire a right to water without applying it to a "beneficial use". Putting aside limited rights to treat instream recreational uses as beneficial ones,[271] this means that someone who anticipates increases in the value of water can exploit his insight only by constructing diversion works and actually committing the water to use. The rule, rather than curbing the speculative impulse, assures that its fulfillment will be accompanied by wasteful economic effort — the premature construction of diversion works or the construction of works that are not economically justified at all.[272]

Further, where the water is nonrenewable groundwater subject to the prior appropriation doctrine, one can expect to see the sort of premature extraction to which the common pool problem gave rise in oil and gas pools. Indeed, the problem may be worse, as potential extractors are not necessarily limited to owners of overlying land (or transferees of their extraction rights), but may include the world at large.

The distortions generated by the rule have implications for regional conflict over water, which we treat here briefly because of its importance as an input to production and transportation of energy resources. First, where two states have access to water, and one is more developed than the other (though the latter may be developing more rapidly or be about to do so), the more developed state may try to apply the water to immediate use and thus secure a valid property right in the water, even though the more efficient solution would be to delay application of the water to any use. Second, if the more developed state makes the appropriation in the less developed one, it thereby effects an uncompensated wealth transfer, for appropriators pay nothing for the water right that their appropriation secures.

All these difficulties come together in a pending case,[273] where the City of El Paso, Texas seeks to make an appropriation of groundwater from an adja-

[271] *See* Tarlock, *Appropriation for Instream Flow Maintenance: A Progress Report on "New" Public Western Water Rights,* 1978 UTAH L. REV. 211 (1978).

[272] *See* C.J. MEYERS & R.A. POSNER, MARKET TRANSFERS OF WATER RIGHTS: TOWARDS AN IMPROVED MARKET IN WATER RESOURCES 39-43 (National Water Commission, Legal Study No. 4, NTIS No. NWC-L-71-009, July 1971). *See also* Williams, *The Requirement of Beneficial Use as a Cause of Waste in Water Resource Development,* 23 NAT. RES. J. 9 (1983); Williams, *Free Trade in Water Resources:* Sporhase v. Nebraska, *ex rel.* Douglas, 2 S. CT. ECON. REV. 173 (1983).

[273] The case is the subject of two reported opinions, City of El Paso v. Reynolds, 563 F. Supp. 379 (D.N.M. 1983) (invalidating a New Mexico statute explicitly forbidding out-of-state water transfers), and City of El Paso v. Reynolds, 597 F. Supp. 694 (D.N.M. 1984) (partially invalidating certain New Mexico legislation that was adopted in response to the first decision and that less crudely burdened interstate transfers). The New Mexico legislature has made still further amendments, *see* 1985 N.M. Laws, ch. 201, and hearings are currently pending on El Paso's applications to the state engineer.

cent region in southeastern New Mexico. As the more developed state, it is in a position to apply the water to immediate use (even though that application may well be premature and even though the more efficient application of the water might ultimately be in New Mexico). The appropriation, if permitted, will transfer a New Mexico resource to Texas without compensation.[274]

State treatment of the export problem, however, typically goes far beyond anything that might be necessary to offset the described market failures. Complete prohibiton of water export is common.[275]

The vulnerability of such restrictions to negative commerce clause attacks may be slightly affected by an old doctrine dealing with wildlife which had been declared by state law to be held in common by the people of the state. In *Geer v. Connecticut*[276] the Court had relied on such nominal public ownership to sustain a statute forbidding anyone to ship out of state any woodcock, ruffled grouse or peacock killed within the state. As to wildlife, the Court subsequently carved out a series of exceptions to the doctrine, and finally overthrew it explicitly in *Hughes v. Oklahoma.*[277] In doing so, it quoted language from a number of intervening opinions and from the dissenting opinion in *Geer*, characterising the theory of state ownership of animals *ferae naturae* as "a fiction expressive in legal shorthand of the importance to its people that a State have power to preserve and regulate the exploitation of an important resource."[278] One of the quoted passages proclaimed:

> A State does not stand in the same position as the owner of a private game preserve and it is pure fantasy to talk of 'owning' wild fish, birds, or animals. Neither the States nor the Federal Government, any more than a hopeful fisherman or hunter, has title to these creatures until they are reduced to possession by skilful capture.[279]

One might more aptly characterise the fiction of "public ownership" in such cases as merely reflecting the facts that the resource is "common to all and the property of none",[280] and that because property rights have been incompletely specified there is unusual justification for state regulation.[281] Certainly it is hard to believe that in 1896, the date of *Geer v. Connecticut*, ruffled grouse were of exceptional importance to the welfare of Connecticut residents.

[274] *See* Post-Trial Brief of Defendant Garza and of Defendant-Intervenors 5, 15 (filed 8 Apr. 1982), strenuously urging the uncompensated nature of the transfer.

[275] *See, e.g.*, MONT. CODE ANN. § 85-1-121 (1983); WYO. REV. STAT. § 41-3-105 (1977), *repealed* 1983 Wyo. Sess. Laws, ch. 167, § 2, *replaced by* § 41-3-115 (1977); NEW MEXICO CODE ANN. § 72-12-19 (*invalidated in* the 1983 *City of El Paso* decision).

[276] 161 U.S. 519 (1896).

[277] 441 U.S. 322 (1979).

[278] *Id.* at 344 (*quoting* Toomer v. Witsell, 334 U.S. 385, 402 (1948)).

[279] *Id.* (*quoting* Douglas v. Seacoast Products, Inc., 431 U.S. 265, 284 (1977)).

[280] *See* United States v. Gerlach Live Stock Co., 339 U.S. 725, 744-45 (1950).

[281] At one time it appeared that in the absence of such a market failure state interference with property rights might violate the due process clause of amend. XIV. *See, e.g.*, Ohio Oil Co. v. Indiana, 177 U.S. 190 (1899).

In any event, in *Sporhase v. Nebraska*[282] the Court recognised that "public ownership" is as much a fiction for water as it is for wildlife. However, apparently because water resources really *are* of critical importance to a state's welfare, the Court indicated that the public ownership fiction could "support a [state's] limited preference for its own citizens in the utilization of the resource."[283] The opinion suggested, however, that any express discrimination against interstate commerce would bear a heavy burden of justification.[284]

It thus appears that with respect to unowned or incompletely owned water resources, the Court will apply at most only a very diluted version of *Reeves*, which the *Sporhase* opinion cited in support of its willingness to allow states more freedom to discriminate as to water than they would normally. Another implication of the opinion, of course, is that with respect to a nonreproducible natural resource, a state can enjoy some of the leeway afforded state-owned conventional enterprises by the *Reeves* doctrine.

2. Monopsony Relation Between Selling as Producer and Buying Pipeline (Coupled with Common Pool Problems)

Natural gas pipelines are a natural monopsony or oligopsony in relation to natural gas producers: because of economies of scale, it is frequently uneconomic to have many pipelines gathering gas from a field, and one pipeline seems common. Under normal conditions, this monopsony could be expected to result in each pipeline's gathering less gas, and paying a lower price, than would be gathered and paid under perfectly competitive conditions.[285]

The monopsony problem in natural gas gathering is exacerbated by common pool factors. An owner who refuses to sell to the monopsonist (or oligopsonist) not only makes no sale, but he finds his interest in the gas sucked out from under him by the owners who have.

Gas producing states have responded to the problem by authorising their regulatory commissions to issue "rateable take" and "common purchaser" orders. The former require the pipeline to take rateably from all producers with which it has made a connection (the take typically to be in proportion to the wells' allowables). The latter go a step further and require the pipeline also to connect with the wells of producers with whom it has no contract, and to take rateably from all. To provide effective protection, a common purchaser

[282] 458 U.S. 941 (1982).

[283] *Id.* at 956.

[284] *Id.* at 957-58.

[285] For a monopsonist the cost of the marginal unit is greater than its price because (assuming he cannot discriminate between suppliers) the purchase of an additional unit increases the price he pays for inframarginal units. Accordingly, he buys less of the item, and at a lower price, than purchasers would under compeitive conditions. *See* E. MANSFIELD, MICROECONOMICS. THEORY AND APPLICATIONS 387-94 (2d ed. 1975).

order must also establish a minimum price which the pipeline is to pay.[286] The regulatory system might thus provide an opportunity for gas producing states (*if* gas were concentrated in few enough states to make concerted action feasible) to force prices ***above*** levels that would prevail under conditions free of monopsony or common pool problems. They could thereby shift wealth from consumers to producers. In contrast with the situation for market demand prorationing, however, we know of no accusations that they have done so.

The Supreme Court in *Cities Service Co. v. Peerless Oil & Co.*[287] upheld a common purchaser order of the Oklahoma regulatory commission against a claim that it violated the negative commerce clause. The Court's opinion suggests a judicial willingness to sustain price floors established by producer states without much scrutiny of the economic reality. (It is, in that respect, similar to the opinion in the *Champlin* case.) The Court did not develop the monopsony or common pool issues. It did, however, note the conclusion of the Oklahoma Corporation Commission that "there was no competitive market for gas in the Guymon-Hugoton Field, that the integrated well and pipe-line owners were able to dictate the prices paid to producers without pipe-line outlets, and that as a result gas was being taken from the field at a price below its economic value."[288] And it referred vaguely to facts of dubious relevance: that "groups connected with the production and transportation of competing fuels complain of the competition of cheap gas"; that the wellhead price is but a small fraction of that paid by domestic consumers at the burner-tip; and that conservation of gas (induced by the higher price) might conform to the national interest.[289]

Had the Court pinpointed the monopsony and common pool issues, one might read the opinion as permitting price floors by a major exporting state only where a serious economic justification was present. Instead, however, it accepted extremely questionable claims. Presumably any state seeking to justify price floors in the future will be able to advance arguments at least as meritorious as the claim that competitors have been objecting to the low price (*i.e.*, to competition).[290]

[286] *See* Meyers, *Federal Preemption and State Conservation in* Northern Natural Gas, 77 HARV. L. REV. 689, 690,694 (1965); Cities Service Co. v. Peerless Oil & Gas Co., 340 U.S. 179 (1950).

[287] 340 U.S. 179 (1950).

[288] *Id.* at 183. At the time, wellhead prices in the field ranged between 3.6 and 5 cents per thousand cubic feet (MCF), although the "commercial heat value" of the gas, in terms of competitive fuel equivalents, exceeded 10 cents per MCF. *Id.* at 181. The Commission set a minimum price of 7 cents per MCF.

[289] 340 U.S. at 187-88.

[290] The case raised no claim of preemption by the Natural Gas Act. For consideration of that issue, *see infra* Ch. III, pp. 109-13.

3. Bilateral Monopoly Relation Between Pipeline as Seller and Distribution Firm as Buyer

As both long-distance natural gas pipelines and local gas distribution companies may have natural monopoly characteristics, the relation between them may be one of bilateral monopoly: one potential seller and one potential buyer, each having no close substitute for the other's business. Such a relationship clearly entails the usual monopoly risk that sales will be at a higher price and of a smaller quantity than would prevail under competitive conditions. Accordingly it is not surprising to see attempts at state regulation.

Such attempts, however, ran foul of the Supreme Court's 1920s view of the negative commerce clause. In *Missouri v. Kansas Natural Gas Co.*,[291] it held that a gas importing state could not regulate the prices at which natural gas originating elsewhere was sold by a pipeline company to a local distribution company. In a parallel case involving interstate sale of electricity, *Public Utilities Commission v. Attleboro Steam & Electric Co.*,[292] the Court denied the equivalent power to the exporting state. Both decisions rested on the now abandoned concept that the validity of state regulation under the negative commerce clause turned on whether the regulated transaction was "interstate" or "local".

It would be hard to reconcile the outcomes in *Kansas Natural Gas* and *Attleboro Steam & Electric* with the Court's conclusion in *Cities Service Co. v. Peerless Oil & Gas Co.*,[293] that producer states were entitled to regulate the wellhead price of gas sold interstate. Whatever nice distinctions one might draw, the controlling fact is the Court's shift, in the years between 1927 and 1950, towards the currently prevailing view: that although state regulations expressly discriminating against interstate commerce are subject to a very heavy (almost insurmountable) burden of justification, regulations not so discriminating are invalid only if "the burden imposed on [interstate] commerce is clearly excessive in relation to putative local benefits."[294]

Serious consequences, however, flowed from the chance event that the state regulatory efforts involved in *Kansas Natural Gas* and *Attleboro Steam & Electric* underwent judicial scrutiny at the time they did. The Court struck down state regulatory efforts for which there was a powerful case in standard economic analysis. So far as states were concerned, then, the Court created a "regulatory gap", and when Congress intervened with the Natural Gas Act in 1938, it justified the intervention as an effort to fill that gap. The resulting developments are discussed below in the section on federal legislative interventions.

[291] 265 U.S. 298 (1924).
[292] 273 U.S. 83 (1927).
[293] *Supra* note 287.
[294] Pike v. Bruce Church, Inc., 397 U.S. 137, 142 (1970).

B. In the European Communities

Inadequate specification of property rights in natural resources is not the major problem that it is in the United States under the Rule of Capture. The reason is that, like most other natural resources, oil and gas — the resources whose fugacious nature creates the problem — are in the Member States of the European Communities either under the ownership, or subject to the exclusive or sovereign rights, of the state.[295] In consequence there is unity of ownership (or at least of control), and common pool problems can only arise if the state grants extraction rights, in the form of licences or concessions, to more than one company or consortium in the same oil or gas deposit. In fact this is a common occurrence, at least offshore, where in order to encourage rapid and thorough exploration and exploitation states have divided their territory on a group pattern, with blocks as small as an *average* of 250 square kilometres in the United Kingdom. Separately licensed blocks often, therefore, overlie a single petroleum deposit. The consequent property right problems can, however, be remedied by the inclusion of compulsory unitisation provisions in the relevant licences or concessions, without the need for regulatory devices of the types encountered in the United States. Such unitisation provisions are to be found in the law of the relevant European Community Member States,[296] but are in fact seldom if ever used, licensees finding it possible to reach agreement among themselves as to the unitisation of their expensive-to-exploit offshore oil fields.

European States with oil resources do not, therefore, have common pool problems as an excuse on which to build a system of controls which may be used in order to capture monopoly returns for themselves or for their licensees. This is not to say that they have refrained from taking control powers which could possibly be exercised with this effect.

In relation to oil and gas, the principal exporting states, the United Kingdom and the Netherlands, both possess powers, particularly over the rate of production of their oil and gas resources, which could be used to restrict production and thereby either put a floor under prices, as we argue has occurred in the United States, or extract other kinds of advantages from other Member States in search of additional secure crude oil supplies. Within the United Kingdom regime[297] two main possibilities exist for the Government: to control the rate of discovery, development and depletion by the size and frequency of allocations of petroleum licences; or, to exercise its powers under the production licences, according to which the licensee may not produce any oil or gas otherwise than in accordance with a development programme approved by

[295] *Supra* Ch. I, pp. 10-15.

[296] *See, e.g.*, Petroleum (Production) Regulations 1982, S.I. 1982 No. 1000, Sch. 5, cl. 25, for the unitisation clause contained in all UK licences; and the Decree of 6 February 1976, *supra* note 54, at Ch. III, art. 28, for the corresponding Netherlands provision.

[297] *See generally* UNITED KINGDOM OIL AND GAS LAW, *supra* note 56.

the Secretary of State for Energy or a production consent given by him.[298] Such programmes and consents may specify the maximum and minimum rates at which production is to be obtained from the field, as well as the equipment and methods of production to be used. The main or ostensible purpose of these controls is to secure safe, efficient production and an appropriate rate of depletion of the resource. It seems unlikely that they would be used for the above-mentioned trade-restricting ends other than in the framework of a Community policy (for example, to maintain a price floor for oil in a time of glut in order to sustain development of alternative energy sources). Nonetheless, the possibility exists, as does the chance that a licensee disadvantaged by the operation of production restrictions for any reason might seek to rely on Community law to invalidate them. We look, therefore, at the constraints which the free trade principles of the EEC Treaty might place on the exercise of these powers.

A general problem raised by the application of article 34 in connection with natural resources is, as already noted, its relevance to controls on production. In this respect the prohibition on measures of equivalent effect to quantitative restrictions on exports may be seen as the cutting edge of a general and fundamental concern, which Paul Reuter articulated in the following terms in the context of the ECSC Treaty:

> The Treaty is primarily concerned with *trade* ... subject to particular exceptions, it contains no dispositions on the conditions of production. Thus it does not prejudge the system of property approved to enterprises.... But must we then admit the legitimacy of national legislation requiring a state authorisation for the creation of new enterprises or the extension of existing installations (commissioning of steel furnaces, for example)? ... It is hardly compatible with the dispositions of the Treaty that a state should exercise the High Authority's function of seeing to the regular supply of the market or of coping with glut. So a state still has the right to refuse a mining concession, but can it refuse all new concessions in a systematic manner?[299]

Contrary to Reuter's expectation, such questions did not quickly present themselves as practical problems within the ECSC framework — or if they did, they were resolved without fuss. Under the general regime of the EEC Treaty, the question cannot arise in quite the way framed by Reuter, because the Treaty does not endow the Commission with the functions of ensuring the orderly supply of the market. Outside the agricultural sector, therefore, the issue is not one of possible conflict between express powers of the Commission and retained powers of the Member States; rather, it is whether the general powers of the Member States in relation to production are cut down by the prohibitions of a Treaty which is primarily concerned with trade and confers

[298] Petroleum (Production) Regulations 1982, S.I. 1982 No. 1000, Sch. 5, cls. 14, 15.

[299] P. Reuter, La Communauté européenne du charbon et de l'acier 180-81 (1953) [our translation].

no central powers over production. Framed in this way, the question seems to demand a negative answer, if only to avoid the creation of a regulatory "gap". In two cases, however, the Court has used language indicating that a production restriction could be contrary to article 34, and in one of them, *Officier van Justitie v. Van Haaster,*[300] it actually found one to be so. The restriction in question was a Dutch national quota on the production of hyacinth bulbs. A similar question arose in *Kramer,*[301] but this time in relation to Dutch catch quotas on sole and plaice. Here we are dealing with fisheries, a resource whose character of exhaustibility bears a closer resemblance to that of mineral resources than of hyacinth bulbs or other agricultural products. The Dutch measures, like the United Kingdom oil depletion powers, were designed for the conservation of that resource. This point was taken by the Court which pointed out that the fact that the measures had the effect, for a short time, of reducing the quantities that states were able to exchange among themselves, did not mean that they were prohibited by the Treaty: the decisive factor was that in the long term such measures were necessary to ensure a steady, optimum yield from fishing.[302]

It is tempting to seek to relate United Kingdom oil depletion policy with these fishery conservation measures by comparing the operation of oil production limitations on the one hand and catch quotas or other fishing limitations on the other. The former's main purpose is to postpone the recovery of part of a quantity which is, within fairly narrow limits, predetermined;[303] the latter, by preventing the recovery of part of a presently existing quantity, to permit the reconstitution of the resource in the future in quantities larger than would have been the case in the absence of the restriction. How would this difference affect the reasoning of the Court in *Kramer*? Such an inquiry might be misconceived. What links *Kramer* and *van Haaster,* and would separate them from a case involving oil depletion controls, is the presence of a Community regulation for the common organisation of the relevant agricultural market. In each case the primary question that the Court appears to have asked itself is whether the national regulation in question was consistent with the Community regime. In each case, it should be noted, that regime itself involved some restriction of production. In *van Haaster* the chosen Community instrument for this purpose was quality controls, and a national production quota was regarded as an additional, and inconsistent restriction. In *Kramer,* on the other hand, the regime itself made provision for conservation measures, including catch quotas, as a possible restriction on production. It was the fact of the in-

300 Case 190/73, [1974] E.C.R. 1123.

301 Joined Cases 3, 4 & 6/76, [1976] E.C.R. 1279.

302 *Id.* at 1313.

303 In question here are "economic" controls on depletion. "Technical" conservation measures, it should be stressed, may be designed to increase ultimate recovery from the reservoir, *e.g.,* by conserving reservoir pressure during extraction, providing for secondary recovery through water or gas reinjection, etc.

consistency of the *van Haaster* measure with the common market organisation which endowed it with its effect of burdening trade, whereas in *Kramer*, a national conservation measure lawfully adopted (because the Council had not yet adopted a Community rule) and of a type envisaged by the regime could not impose such a burden.

After the *Kramer* case, therefore, one was left in some doubt as to the compatibility of a production restriction with article 34 in the absence of a market organisation. That doubt has been resolved by the *Groenveld* and *Oebel* cases which, as we have already seen,[304] announce an approach by the Court to article 34 which admits restrictions applying alike to production for home and export markets. Such is certainly the case with the instruments of oil depletion policy, whatever the difficulties it creates for other Member States of the Community. Within the EEC framework, therefore, the answer to Reuter's ultimate question is to be found not in article 34, but in the development of centralised Community powers of control akin to those provided in the agricultural sector. We look in the next chapter at the possibilities of this type offered by the Treaty.

[304] *Supra* pp. 37-38.

Chapter Three

Centralised Energy Policy

I. The United States

The main subject of this section will be national policy relating to the two energy sources — natural gas and crude oil — on which we focussed in the prior section. We offer no detailed consideration of national policy for coal and nuclear fuels because the bases of federal power in these sectors are simply not present in Europe. As to coal supply, if we put aside federal environmental policy,[1] the federal role derives from the Federal Government's ownership in place of about half of United States coal reserves. In the Western United States some 60 percent of reserves are in federal ownership, and because of land ownership patterns federal leasing policy will as a practical matter affect over 80 percent of Western reserves.[2] While federal coal has historically represented a relatively small proportion of the coal in production, the federal fraction of production has been rising steadily (to over 10 percent in 1981) and this trend

[1] Federal environmental policy is, in fact, critical to coal development. The Surface Mining Control and Reclamation Act of 1977, P.L. No. 95-87, 91 Stat. 448 (codified as amended at 30 U.S.C. paras. 1201-1328 (Supp. IV 1980)), seeks to diminish the adverse environmental effects of strip mining. Environmental law has also dramatically influenced coal development indirectly, by virtue of the 1977 amendments to the Clean Air Act, P.L. No. 95-95, 91 Stat. 685 (codified at 42 U.S.C. paras. 7401 *et seq.*). These amendments permitted the Environmental Protection Agency (EPA), and perhaps forced it, to require that new fossil-fuel burning power plants emit into the air no more than some percentage of the sulphur *originally contained in the fuel* burned. The effect of these restrictions was to deny low-sulphur western coal its natural competitive advantage with regard to controlling air pollution: new plants can comply with the regulations only by means of expensive flue-gas desulphurisation equipment, and plants with such equipment can achieve compliance with the requirements even when they burn high-sulphur eastern coal. *See* Ackerman & Hassler, *Beyond the New Deal: Coal and the Clean Air Act*, 89 YALE L.J. 1466 (1980); B. ACKERMAN & W. HASSLER, CLEAN COAL/DIRTY AIR (1981).

[2] For an estimate of federal coal ownership, *see* H.R. REP. No. 681, 94th Cong., 1st Sess. 9, *reprinted in* 1976 U.S. CODE CONG. & AD. NEWS 1943, 1945.

seems likely to continue.[3] Federal power with respect to such coal is based on the Property Clause of the Constitution,[4] and is very broad indeed.[5]

In nuclear fuel supply, the Federal Government not only owns a large proportion of uranium reserves but also possesses a commanding lead over the states in the expertise necessary for the effective regulation of private supplies. Federal domination of the nuclear industry originated with the military uses of atomic power in World War II and continued thereafter. On its creation in 1946, the Atomic Energy Commission (AEC) became the owner of all fissionable materials and all facilities that used or produced such materials, as well as exercising control over use of related technical information. In 1954 this federal control was loosened in the direction of admitting private industry as an owner of nuclear facilities and user of fissionable materials (owned by the AEC but licensed to private firms).[6] In European nuclear powers, such as Britain and France, a similar, though not identical post-war evolution occurred; but this led to a *Member State* dominance of the nuclear industry which, as we shall see, integration and centralisation attempts at the European level have hardly weakened. In the United States, apart from an occasional venture into atomic entrepreneurship, as by the New York State Nuclear Authority, state involvement has been largely limited to environmental regulation; even there, because of a high degree of express federal preemption, state action has been relatively peripheral.[7]

Natural gas and crude oil account for nearly 70 percent of America's primary energy consumption (natural gas, 24 percent; petroleum, 42 percent).[8] Perhaps because of that dominance, they are the sources with respect to which federal policy has had the most dramatic impact. The federal intervention with respect to natural gas is, moreover, closely related to the state interventions, for it was the restriction of those interventions by the federal judiciary that triggered federal legislation.

The federal interventions are striking in two respects. First, they manifest a striking tendency to impede the free flow of commerce among the states. Second, they seem impossible to ascribe in significant part to any national

[3] *See* Nelson, *Undue Diligence: The Mine-It-or-Lose-It Rule for Federal Coal*, 1983 Regulation 34 (Jan./Feb. issue). For an excellent treatment of the subject generally, *see* R.H. Nelson, The Making of Federal Coal Policy (1983).

[4] Art. IV, § 3, cl. 2.

[5] *See generally* Kleppe v. New Mexico, 426 U.S. 529 (1976).

[6] *See, e.g.*, E. Rolph, Nuclear Power and the Public Safety 21-22, 27 (1979).

[7] Pacific Gas & Elec. Co. v. State Energy & Resources Cons. & Dev. Comm'n, 461 U.S. 190 (1983), however, may signal a sharp decrease in the scope of federal preemption as understood by the courts. *See also* Silkwood v. Kerr-McGee Corp., 464 U.S. 238 (1983); *but see* County of Suffolk v. Long Island Lighting Co., 728 F.2d 52 (2d Cir. 1984).

[8] *See* Energy Information Administration, Annual Energy Review 1984. The figures are for 1984. The remainder is made up by coal (23%), hydropower (5%), and nuclear power (5%). The total adds to less than 100% due to rounding.

purpose other than a redistributive one. In the conclusion, we will consider whether this picture may have general implications for the ability of a federal system to protect and enhance free trade among its units.

A. Federal Regulation of Wellhead Prices of Natural Gas

State and federal interventions in the pricing of natural gas provide an opportunity to study the relation between state gambits, responses to them by the federal judiciary, and responses to the latter by Congress. However, as historical accidents played a role in critical developments, it is impossible to draw any clear object lesson. We will trace the chain of events by which federal law came to regulate wellhead prices of natural gas, seek out the purposes behind continued regulation, and evaluate the consequences for United States energy policy and the implications for its allocations of power. Finally, we will close by treating the aftermath of the mid-1970s gas shortage and the status of state conservation legislation in the face of federal intervention.

1. Origins

Supreme Court decisions in the mid-1920s, *Missouri v. Kansas Natural Gas Co.* and *Public Utilities Commission v. Attleboro Steam & Electric Co.*,[9] created a roadblock to state efforts to intervene in the bilateral monopoly relation between interstate suppliers of natural gas and electricity, on the one hand, and local distribution companies on the other. The first decision invalidated the efforts of the *importing* state to control prices charged by an interstate pipeline for imported gas sold wholesale to such a distribution company, and the latter struck down comparable regulatory efforts, in relation to electricity, by the *exporting* state. Assuming that the same legal principles governed natural gas as governed electricity, the two decisions appeared to prevent *any* state from regulating the wholesale prices of natural gas moving in interstate commerce. Congress perceived this as a "regulatory gap", and filled it with the Natural Gas Act of 1938. Said the House Report in support of the bill:[10]

> The States have, of course, for many years regulated sales of natural gas to consumers in intrastate transactions. The States have also been able to regulate sales to consumers even though such sales are in interstate commerce, such sales being considered local in character and in the absence of Congressional prohibition subject to State regulation. ... There is no intention in enacting the present legislation to disturb the States in their exercise of such jurisdiction. However, in the case of sales for resale, or so-called wholesale sales, in interstate commerce (for example, sales by producing companies to distributing companies) the legal situation is different. Such transactions have been con-

[9] *Kansas Natural Gas Co.*, 265 U.S. 298 (1924); *Attleboro Steam & Electric Co.*, 273 U.S. 83 (1927).

[10] H.R. Rep. No. 709, 75th Cong., 1st Sess. 1-2. Natural Gas Act of 1938, 52 Stat. 821 (codified at 15 U.S.C. § 717 *et seq.*).

sidered to be not local in character and, even in the absence of Congressional action, not subject to State regulation. (*See Missouri v. Kansas Gas Co.* (1924), 265 U.S. 298, and *Public Service Commission v. Attleboro Steam & Electric Co.* (1927), 273 U.S. 83). The basic purpose of the present legislation is to occupy this field in which the Supreme Court has held that the States may not act.

Section 1(b) of the Act authorised the Federal Power Commission (FPC) to regulate "the sale in interstate commerce of natural gas for resale", and thus clearly filled the perceived "gap". The FPC has regulated pipelines' prices to local distributors ever since. The propriety of such regulation has not occasioned major political controversy.[11]

The language of Section 1(b) was obviously also applicable, if read literally, to sales at the wellhead by independent natural gas producers to interstate pipelines. As we have seen, these sales involved a radically different problem. The monopsony power of the pipelines, and the "common pool" relation between gas producers, were likely to result in selling prices *below* those that would occur in a purely competitive market where producers had conventional fully-protected property rights in their supplies prior to extraction. States had responded to that difficulty with "rateable take" and "common purchaser" orders. As we have seen, their authority to do so was ultimately — in 1950 — to be upheld by the Supreme Court in *Cities Service Co. v. Peerless Oil & Gas Co.*[12]

The legislative history of the Natural Gas Act provided no affirmative evidence of any intent to regulate such wellhead prices. The issue was not a subject of discussion in the Congressional consideration of the Act.[13] A report of

[11] The widespread acceptance of the regulation of pipelines has perhaps been due to the assumption that they constitute a natural monopoly. Recent research has tended to suggest that in fact for a large proportion of gas sold, there are enough potential suppliers, by virtue of parallel lines, to provide a substantial degree of competition. *See* N. Clark & G. Clark, Governments, Markets and Gas (1984); American Gas Association, Competitition in the Natural Gas Industry (14 Feb. 1984). There is already concern that the present form of regulation may, by blunting the transmission of market signals from the burner-tip back to the producer, partially account for the fact that a glut has developed since 1981 even as prices were rising. The addition of new evidence as to the competitive potential of the industry may undermine the former consensus.

[12] 340 U.S. 179 (1950). *See supra* Ch. II, text accompanying notes 287-90.

[13] *See* Note, *Federal Price Control of Natural Gas Sold to Interstate Pipelines,* 59 Yale L.J. 1468, 1479 (1950) [hereinafter cited as *Yale Note*].

Prevalence of the assumption in Congress that the Act did not cover wellhead prices can be found in the fact that Congress included, at the behest of a "gas-state" senator, a provision requiring that the Federal Power Commission (FPC) set its rates at "the lowest posssible lawful rate consistent with the maintenance of adequate service in the public interest." *See* M.E. Sanders, The Regulation of Natural Gas: Policy and Politics, 1938-78, at 52 (1981). (The provision was deleted in 1942. *Id.* at 214 n.17.) This requirement made sense for states that produced natural gas and were anxious that coal-producing states might induce the FPC to set pipeline

the Federal Trade Commission, analysing the issues posed by the natural gas pipelines and constituting a key ingredient of the background history, had identified the radical difference between sales by the pipeline to local distribution companies (the area where the *Kansas Natural Gas Co.* decision created a "gap") and sales by independent producers to the pipelines. As to the latter, it recognised the dilemma of the independent producers. Its recommendation was a federal law *reinforcing* the state regulatory efforts, namely a federal "hot-gas" statute providing federal remedies against anyone who transported in interstate commerce gas purchased in violation of the state regulatory efforts.[14]

Section 1(b), however, provided literally for FPC regulation of independents' wellhead prices, so opponents of such regulation were reduced to arguing that those prices were exempted from federal regulation pursuant to a clause exempting "the production and gathering of natural gas". In 1954, however, the Supreme Court, in *Phillips Petroleum Co. v. Wisconsin*, construed that exception as being limited to activities occurring before the sale to the pipeline.[15] Dissenting, Justice Douglas pointed out prophetically that the prices paid to the independent producer determine his profits,[16] and "his profits and the profits of all the other gatherers, whose gas moves into the interstate pipelines, have profound effects on the rate of production, the methods of production, the old wells that are continued in production, the new ones explored, etc."

An irony of the case is that by the time of this critical extension of FPC jurisdiction, the Court had already abandoned the doctrines creating the alleged regulatory gap. Its 1950 decision in *Cities Service Co. v. Peerless Oil & Gas Co.*[17] had not only approved the "common purchaser" order at stake there, but also reflected a far less constraining attitude toward state regulatory efforts — an attitude which almost surely would have led to approval of the state regulatory gambits reviewed in *Kansas Natural Gas* and *Attleboro Steam & Electric*.

2. Purposes of Continued Regulation of Independents' Wellhead Prices

Since the extension of federal price regulation to independents' wellhead prices arose as a result of Congress' negligent use of overbroad language and the Supreme Court's literalism, a search for any initial Congressional purpose would be unavailing. One cannot ascertain a purpose for an act that was never

charges so high as to price natural gas out of the market. Of course, it would make no sense for a representative of a state producing natural gas in substantial quantities to propose the requirement if he supposed that FPC jurisdiction also covered wellhead prices.

[14] *Yale Note, supra* note 13, at 1479 and 1479 n.73.

[15] 347 U.S. 672, 678 (1954).

[16] *Id.* at 690.

[17] 340 U.S. 179 (1950).

intended. After the intitial confusion, however, there were efforts in Congress to clarify the Act by expressly precluding federal regulation of such prices, and they successively foundered. Thus, while it was *accident* that the Act shifted the political burden onto opponents of regulation, the conduct of regulation's proponents in frustrating deregulation efforts was *intentional.*

The intent can be fairly clearly identified as the capture and redistribution of economic rents. Although proponents of regulation occasionally talked of monopoly power among the independent producers,[18] evidence of lack of concentration contradicted these suggestions.[19] The consensus among economists has been that the market was workably competitive.[20] The second theme sounded by proponents of regulation was that of "windfall profits". President Truman, for example, vetoing a 1950 bill that would have exempted the sales by independents to interstate pipelines, expressed his desire "to prevent unreasonable and excessive prices, which would give large windfall profits to gas producers, at the expense of consumers."[21]

Rapidly increasing field prices and producer profits did suggest the presence of large rents.[22] The post-World War II surge in demand for natural gas provided economic incentives to search for more supplies, which were more costly to find and develop than the gas discovered earlier. Holders of old reserves naturally enjoyed inventory profits.[23] The increased competition for supplies among pipelines also undercut their earlier monopsony power and thus tended to correct the artificial depression in prices that had resulted from that monopsony.[24] The evidence of large rents, coupled with the rhetoric of "windfall profits" and the lack of evidence of monopoly, suggest that rent redistribution was necessarily the purpose of those who blocked repeal of wellhead price controls.

3. Consequences and Implications for the Federal Allocation of Powers

Breyer has aptly identified the FPC's use of conventional cost-of-service rate-fixing for the purposes of transferring economic rents to consumers as regulatory "mismatch".[25] After a period of floundering with firm-by-firm rate-fixing, the Commission switched to rate-making by larger units — first for each producing *area* and ultimately for the entire nation. Recognising that

18 *See* Public Papers of Harry S Truman 1950, at 257 (1956) (message explaining veto of bill to exempt wellhead prices of natural gas sold by independents).
19 *See Yale Note, supra* note 13, at 1492-93.
20 *See* P. McAvoy, Price Formation in Natural Gas Fields (1953); J. Mulholland, Economic Structure and Behavior in the Natural Gas Production Industry (Federal Trade Commission, 1979); S. Breyer, Regulation and Its Reform 242 (1982).
21 Public Papers of Harry S Truman 1950, *supra* note 18, at 258.
22 *See Yale Note, supra* note 13, at 1484-1500.
23 *See* S. Breyer, *supra* note 20, at 243.
24 *Id.*
25 *Id.* at 240-60.

a single price would either fail to capture rents, or would have drastic adverse impacts on exploration and development of new, higher-cost supplies, it moved to a multi-tier system. But a necessary consequence of that decision was that at the prices fixed for old, low-cost gas, demand outran supply. Moreover, as FPC jurisdiction clearly did not extend to intrastate sales of gas, producers of new gas could escape regulation by keeping their supplies in the intrastate market. Finally, the Commission seems to have simply made some unluckily low guesses as to the costs of new supplies.[26]

As a result, net additions to natural gas reserves had by the early 1970s fallen dramatically. The fraction of those reserves dedicated by natural gas producers to the interstate market (*i.e.*, committed under FPC procedures to sale outside the state of origin) had fallen even more sharply (in 1970 and 1972, for example, to zero).[27] By the winter of 1976-1977 there was a supply curtailment of about 20 percent.[28] The impact of the shortage was quite arbitrarily distributed, depending initially on which pipeline served a particular state and the security of that pipeline's sources of supply.[29] The FPC attempted to allocate the shortage by means of eight categories of priorities,[30] which were theoretically absolute: no user in the seventh category, say, should have his supplies reduced until all users in the eighth category had been cut off altogether.[31] Even had the principle been enforced, the categories represented only a very crude stab at mimicking the market, which would force customers to curtail their gas uses in a least-cost sequence (*i.e.*, curtailing first those uses for which substitutes were cheapest). In any event, the clamour of disappointed victims led the Commission to establish a procedure for pipeline-by-pipeline modification of the curtailment principles, and priority allocation was partially compromised by a principle of "equal" treatment, *i.e.*, pro rata reduction from historic consumption levels for all established purchasers.[32] One class of customer, however, enjoyed virtually absolute protection from cutbacks: the established residential consumer.

Such allocations, of course, directly frustrate the efficiency goals of free trade. A necessary condition for the efficient allocation of any commodity is that the value of the marginal unit must be the same for each user.[33] If the value

[26] *Id.* at 250-53.

[27] *See* American Public Gas Ass'n v. FERC, 587 F.2d 1089, 1095 n.5 (D.C. Cir. 1978).

[28] S. BREYER, *supra* note 20, at 244.

[29] *Id.* at 245.

[30] *Id.* at 254.

[31] Federal Power Commission Order No. 467, 49 F.P.C. 85, 87-88 (1973).

[32] *See* S. BREYER, *supra* note 20, at 258.

[33] Expressed slightly more precisely, the condition requires that the marginal rate of substitution (MRS) between any two goods, *a* and *b*, must be the same for any two consumers, *A* and *B*, having both goods:

$$\text{MRS}^{A}_{a,b} = \text{MRS}^{B}_{a,b}$$

See H. KOHLER, WELFARE AND PLANNING 8-11 (1966).

to user A of his marginal thousand cubic feet ("MCF") of natural gas is $1.00, and the value to user B of his marginal MCF is $2.00, obviously a system under which user A was enabled to relinquish his marginal unit to B could enhance the welfare of both parties. Freezing the users' historical consumption pattern and curtailing it pro rata prevented such reallocations. And even allocation by priority did not correspond with marginal value.

The bifurcation of the intrastate and interstate markets, and the cessation of new dedications of gas for sales across state lines, compounded the problem and heightened the irony. As to that particular effect, not only did the country lose the advantages of reallocation in response to price signals, but state lines became the formal instrument for defining the barriers.

It is hard to believe that state parochialism, free of any constraints imposed by the Supreme Court or by Congress, could have brought about such prolonged and dramatic impediments to free trade. Producer states enacting legislation seeking to raise prices materially above market levels would have lost their sales; consumer states imposing price limits below market levels would have lost their supplies. Efforts by producer states to hoard their supplies, as in *Oklahoma v. Kansas Natural Gas Co.*[34] and *Pennsylvania v. West Virginia*,[35] would have inflicted severe opportunity costs on producing companies, which would probably have led them to seek — with some prospect of success — a reversal of state policy.

Natural gas pricing developments, therefore, seem to provide significant support for Kitch's thesis that federal law has by no means been the friend of free national trade,[36] and some support for his thesis that the Supreme Court's interventions in the name of free interstate trade have actually tended to hinder that trade.[37]

The second thesis is, however, overdrawn. As we have seen, the causal connection between the Supreme Court's restrictive decisions in *Kansas Natural Gas* and *Attleboro Steam & Electric*, on the one hand, and the federal regulation of independents' sales to interstate pipelines, on the other, is weak. Intervening causes — sloppy statutory drafting by Congress and literal-minded interpretation by the Supreme Court — played a critical role.[38] Finally, widespread political support for the capture and redistribution of economic rents, coupled with either a political incapacity to anticipate the efficiency costs of price regulation, or an indifference to those costs, thwarted any legislative

[34] 221 U.S. 229 (1911).

[35] 262 U.S. 553 (1923).

[36] *See* Kitch, *Regulation and the American Common Market*, in REGULATION, FEDERALISM, AND INTERSTATE COMMERCE 10 (A.D. Tarlock ed. 1981).

[37] *Id.* at 46; *see also id.* at 43.

[38] Note also that 11 years passed between the decision in *Attleboro* and the enactment of the Natural Gas Act in 1938. Intervening Congresses perhaps viewed *Attleboro* as creating not a gap but an oasis. Further, note that Congress could have resolved the problem merely by reversing the offending decisions, *i.e.*, by validating the state regulatory efforts that *Kansas Natural Gas* and *Attleboro* had struck down.

cure. In the area of crude oil pricing, as we shall see, these latter forces were sufficient to produce a very similar destruction of free interstate trade, at a time when prevailing Supreme Court doctrine created no material "regulatory gap" at all.

A recurrent theme of this study is the ability of federal institutions to handle the relation between national and local interest. So a natural question is the role of sectional self-interest in bringing about the mid-1970's gas shortage. Some studies of the issue have found that factor to be of surprisingly modest importance. Mitchell studied a number of crucial deregulation votes in the House of Representatives in the period when the shortage was apparent to everyone, and found ideology to be the most critical explanatory variable. He reasoned that the owners of gas-heated homes, who enjoyed the low prices and were virtually certain not to have their supplies curtailed, were the prime beneficiaries of regulation. He identified residents of states suffering very high curtailments of supply as regulation's prime victims. If votes were cast in accordance with sectional interest, then one could expect representatives of states having a high percentage of gas-heated homes to oppose deregulation, and representatives of states with high percentages of curtailment to favour it. But in fact, the votes showed little correlation with such interests. By contrast, ideology, as measured by each representative's rating by the ADA (Americans for Democratic Action), correlated with the actual votes extemely well.[39]

This portion of Mitchell's analysis addresses only the distribution of votes as among the consumer states. But one would expect at least that comparison between the producer and consumer states would reveal a powerful role for regional self-interest. Ideology, however, provided at least as close a fit. Mitchell applied his ideological model to the 42 representatives of the major gas-producing states (Texas, Oklahoma, Louisiana, Kansas and New Mexico); it predicted 39 votes for deregulation and 3 votes against. In fact the region produced 40 votes for deregulation — the 39 individuals predicted by the ideological model, plus one switch.[40]

It might be wrong, then, to characterise the mid-1970's natural gas shortage as a failure of the federal system to handle sectional conflict. It more accurately appears as a failure of the political system to reconcile efficiency values with

[39] *See* Mitchell, *The Basis of Congressional Energy Policy,* 57 TEX. L. REV. 591 (1979).

[40] *Id.* at 601. Studies reaching results similar to Mitchell's are reported in J.P. KALT, THE ECONOMICS AND POLITICS OF OIL PRICE REGULATION 244-46 (1981).

Mitchell's study has been criticised for its use of statewide data on residential energy use, supply curtailment and gas production to analyse the decisions of Representatives (who are normally elected on a district basis rather than a statewide basis) and for its lack of explicit focus on the role of coal-producing states. *See* M.E. SANDERS, *supra* note 13, at 208 n.31, 121-23, 158-60. The weight of these criticisms is unclear, however, and in any event Sanders' own data show that party affiliation (presumably a rough proxy for ideological commitment) had great explanatory power with respect to critical votes. *See id.* at 160.

popular zeal for the capture and redistribution of rents. (A far less crude device for reconciling those purposes — deregulation of prices and imposition of a "windfall" profit tax — lies ready to hand.[41] The shortage does, however, raise the question whether the availability of equally broad power in the national legislature to address such issues leads to better or worse results than would have unfolded if the federal system had allowed each state equally broad power to pursue its own interest. State power to beggar one's neighbour is naturally limited by competition among states (*e.g.*, a producing state's ability to secure cartel profits is great only if the price elasticity of demand for its product is low, and that is unlikely to be the case if there are many producing states). On the other hand, it is not constrained by a direct political need to accommodate out-of-state interests (as a condition of enacting legislation). The limits on national power are a mirror image of this pattern. The natural gas story provides no clear evidence that either is to be preferred.

4. Sequels

There are two sequels to the natural gas story: (1) the passing of the crisis and (2) the impact of federal regulation on the capacity of states to remedy the problems caused by the Rule of Capture as applied to fugacious resources such as oil and gas.

(a) Passing of the Mid-1970's Natural Gas Crisis

The gap between supply and demand eroded gradually for a variety of reasons: (1) regulatory measures under the Natural Gas Act reduced the price differential between intrastate and interstate gas; (2) industry switched from natural gas to oil in response to price increases and in order to assure more secure supplies (ironically in view of expressed national concern about the security of Middle Eastern energy resources); (3) residential consumers cut back consumption in reponse to the price increases; and (4) imports from Canada and Mexico rose.[42] The Natural Gas Policy Act of 1978 (the "NGPA")[43] carried these developments further. It made price control nationwide, thus correcting the Natural Gas Act's adverse effect on interstate sales, and replaced the FPC's general authority to set "just and reasonable" rates[44] with a complex set of natural gas vintages, each with its own price ceiling and rules for escalation. The prices of most new vintages were set at levels roughly similar to those then prevailing in the world market for the energy equivalent in oil (about $12 per barrel). And deregulation of some types of gas, especially "deep" gas pro-

[41] For a discussion of the superiority of taxes over price control as a device for recapture and redistribution of rents, *see* S. Breyer, *supra* note 20, at 240-60; *see also* Williams, *Energy Policy in the Cold Light of Morning*, 61 Tex. L. Rev. 571 (1982).

[42] *See* S. Breyer, *supra* note 20, at 259.

[43] P.L. No. 95-621, 92 Stat. 3352 (codified at 15 U.S.C. paras. 3301 *et seq.*).

[44] 15 U.S.C. para. 717c.

duced from below 15,000 feet, provided additional leeway to bring supply and demand into alignment.

Apart from its general tendency to capture economic rents in the producer states, the NGPA's modifications of federal regulations have had an important additional effect on the regional distribution of power and wealth. Regions served by pipelines with comparatively large supplies of "old" gas subject to low price ceilings (primarily interstate gas) now enjoy an artificial competitive advantage over regions dependent on supplies with little or no gas of that sort. Firms' decisions on location, made in response to this advantage, will over time tend to erode it,[45] as will the gradual exhaustion of the supplies of gas subject to these low ceilings. For the intermediate term (5-10 years), however, it may act as a significant irritant in the on-going regional conflicts over energy.

(b) Effect of National Action on State Power to Correct Market Failures

Congressional action, such as the Natural Gas Act and the NGPA, alters the legal environment in which courts consider the validity of state legislation, giving challengers an opportunity to claim that the federal action has preempted it, either because of express or implied conflict.[46] And an implied conflict may be found in Congressional intent to "occupy" a field of legislation.[47]

A key decision, *Northern Natural Gas Co. v. Corporation Commission*,[48] exposes even legitimate state conservation laws to the risk of preemption by the Natural Gas Act.[49] The case deals with state efforts to handle the familiar problem of pipeline monopsony (or oligopsony) and the common pool. Northern had a contract with Republic Natural Gas Company requiring it to purchase

[45] *See* Federal Energy Regulatory Commission, Notice of Inquiry, Impact of the NGPA on Current and Projected Natural Gas Markets, Docket No. RM82-26-000, at 21-24 (28 Apr. 1982); M. Russell, Natural Gas Deregulation: Overview of Policy Issues 20-24 (Resources for the Future Discussion Paper D-92, Washington, D.C., Apr. 1982).

[46] The classic formulation is that of Hines v. Davidowitz, 312 U.S. 52, 67 (1941), declaring that the issue is whether the state law "stands as an obstacle to the accomplishment and execution of the full purposes and objectives of Congress."

[47] Occupation of the field is sometimes described as a basis for preemption independent of "actual conflict" with the federal statute. *See, e.g.*, L. Tribe, American Constitutional Law paras. 6-24 & 6-25 (1978). The distinction seems overrefined. It seems to mean little more than the obvious point that conflict need not take the form of the state's requiring X while federal law requires non-X, which would put a party subject to both laws in an impossible bind. A conflict may occur even though objects of the legislation could comply with both. For example, when the federal statute represents a determination that it achieves a correct balance between regulatory benefits and costs, important values would be sacrificed (for insufficient return) if its objects were compelled to comply with state laws on the subject as well. *See, e.g.*, City of Burbank v. Lockheed Air Terminal, Inc., 411 U.S. 624 (1973).

[48] 372 U.S. 84 (1963).

[49] Compare the effect of the negative commerce clause on such laws, *supra* Ch. II, text accompanying notes 258-70.

all the gas that Republic produced from the Kansas portion of the Hugoton field, within the allowables set by the state regulatory agency, the Corporation Commission. Northern also had contracts with other producers in the field, under which its purchase obligations were expressly subject to the Republic contract; *i.e.*, under those contracts, if the aggregate of the allowables exceeded Northern's demand for gas, Northern was entitled to inflict the entire burden of its purchase cutbacks upon the non-Republic producers. When the Corporation Commission increased the Hugoton allowables in 1958,[50] Northern in fact chose that course. Since the result would have been to enable Republic to drain gas away from the land of the other producers, the Corporation Commission responded with a "rateable take" order, requiring that Northern (and other purchasers) take gas from all wells in proportion to their allowables. Northern challenged the order as preempted by the Natural Gas Act, and the Court sustained the challenge.

Two themes run through the decision. First was concern that the order bore directly on *purchasers* of the gas, rather than on producers.[51] The Court appeared to think that the order imposed significant bookkeeping obligations on the purchasers in the effort to balance production runs and allowables at the various wells.[52] Yet it appears that Northern would have to gather the necessary information for its own purposes in any event, so that compliance would require no serious extra administrative burdens.[53] By contrast, for the Corporation Commission to enforce the order by itself controlling the producers would require it to take on this task. More important, in some situations — where there is more than one purchaser in a field — protection of correlative rights and maintenance of efficient levels of production would be difficult (if not impossible) to achieve by means of orders directed only at producers.[54] Thus, depending upon how serious the Court was about freeing purchasers from such administrative detail, the decision carried a serious risk to state protection of correlative rights.

The second theme was concern that the regulation would have an effect on natural gas prices.[55] But all state conservation regulation has price implications. The setting of allowables limits what a purchaser can get in one field and thus may drive him to another, with higher prices.[56] Indeed, as conservation regulation is aimed at solving the "common pool" problem that would otherwise exist — each owner producing more rapidly than the optimal rate because of his

[50] It seems likely that it did so in order to improve Kansas' share of production in the Hugoton field, at the expense of Texas and Oklahoma. *See* Meyers, *Federal Preemption and State Conservation in* Northern Natural Gas, 77 HARV. L. REV. 689, 698-99 (1964).

[51] 372 U.S. 84, 92, 94 (1963).

[52] *Id.* at 92.

[53] *See* Meyers, *supra* note 50, at 691.

[54] *See id.* at 693-94.

[55] 372 U.S. 84, 91, 92 (1963).

[56] *See* Meyers, *supra* note 50, at 697.

disregard of effects on other owners in the pool — its basic effect, presumably, is to raise the price of oil or gas in the short run. Yet the opinion did nothing to draw a line between permissible and impermissible impacts on price.

One would expect that the sequel to such a decision would be a series of further decisions, elaborating just what administrative burdens on purchasers, and what price effects, state conservation regulation might inflict consistent with the Natural Gas Act. No such decisions occurred prior to adoption of the NGPA. Appellate decisions on Natural Gas Act preemption do not reveal why. Students of state gas regulation appear to believe that the conservation agencies of the producing states, especially Texas, have failed to develop an effective gas prorationing system, perhaps in part as a result of inhibitions caused by the *Northern Natural* decision.[57] For the *Northern Natural* doctrine, that failure appears to have aborted the usual process of testing a case's limits.

The extent to which federal regulation of natural gas may currently preempt state conservation legislation is most unclear.[58] The purposes of the regulatory

[57] *See* D. PRINDLE, PETROLEUM POLICIES AND THE TEXAS RAILROAD COMMISION 97-107 (1981).

[58] Preemption cases outside the area of conservation regulation have not been particularly startling or revealing. In Maryland v. Louisiana, 451 U.S. 725 (1981), several consumer states and pipelines sought relief from Maryland's "First-Use Tax". Besides imposing the tax (in effect as a toll for transit of federal offshore natural gas through Maryland), the statute also purported to establish its economic incidence, requiring the pipeline to pass the tax forward to the purchaser of the gas. (Under free market conditions, the tax would in the long run probably fall primarily upon owners of the gas — here the federal government. As Louisiana offshore gas is indistinguishable from any other, customers would probably switch to other suppliers rather than pay a higher price than they would have been charged in the absence of the tax. Parties bidding for leases of the gas would reduce their bids in anticipation of the tax.) The effect was to raise a potential conflict with the power of FERC under the NGPA, for the latter, besides establishing a price for offshore gas, had also conferred authority on FERC to permit certain expenses to be added onto that cost. Paragraph 110 of the NGPA (15 U.S.C. para. 3320), provided that the price for a first sale of natural gas could exceed the statutory ceiling to the extent necessary to recover various costs, including costs of transportation, if borne by the seller and "allowed" by FERC. As FERC had in fact allowed the tax to be added onto the seller's price, the conflict was only potential. The actual finding of preemption seems to do little more than state the obvious: in the event of any conflict, the federal directive would necessarily control, so that there was no situation in which the state directive could have any independent force.

At least two states have adopted statutes requiring pipelines to furnish natural gas to owners of land through which the pipeline runs (Oklahoma) or within one-half mile of the pipeline (Kentucky). As such provisions would leach away gas dedicated to interstate sale, it is not surprising that two circuit courts of appeal have held them preempted by the NGPA. *See* Backus v. Panhandle Eastern Pipeline Co., 558 F.2d 1373 (10th Cir. 1977); Public Service Co. of Kentucky v. FERC, 610 F.2d 439 (6th Cir. 1979). *But cf.* Boggs v. McDonald, No. 81-1098 (4th Cir., 6 Oct. 1981).

scheme are now radically more complex than those which the Court imputed to Congress at the time of *Northern Natural.* In that era, the Court identified the regulatory purposes as being "to afford consumers a complete, permanent and effective bond of protection from excessive rates and charges",[59] and "to protect consumers against exploitation at the hands of natural gas companies".[60] So far as wellhead prices are concerned, however, the NGPA has provided for the deregulation of "new" gas prices according to a fixed schedule.[61] Exhaustion of "old" gas, which in the nature of things is bound to occur relatively soon, will effectively terminate all wellhead controls. Since the Act sets in progress the restoration of a free market at the wellhead, it seems fair to infer a Congressional purpose to do just that. State conservation laws may be assessed in light of their conformity to this overriding Congressional purpose.

For rateable take orders — the state intervention that precipitated *Northern Natural* — validity might thus turn on whether the order can be seen as merely trying to correct the problems of oligopsony and correlative rights that distort the wellhead market. A recent Mississippi decision[62] upheld a state order that (as modified by the state court) appeared reasonably well calculated to mimic an undistorted market; in providing that the pipeline must take from producers with whom it had no contract, it required only that it offer "reasonable terms, including a reasonable price, determined by reference to prevailing market conditions and other appropriate economic considerations."[63] The Court's decision did not, however, rest on any such precise effort to find congruence between the federal and state purposes.

For market-demand prorationing, the test suggested above — upholding those interventions calculated to mimic an undistorted market — would seem

The only significant decision against preemption has been the holding in Mobil Oil Corp. v. FPC, 463 F.2d 256 (D.C. Cir.), *cert. denied,* 404 U.S. 976 (1972), that state court interpretation of the royalty clauses in oil and gas leases is — at least as a general matter — not preempted. *See id.* at 265-66 for possible qualifications. As such leases antedate the dedication to interstate commerce, and the clause merely provides for computation of landowner reimbursement for the right to extract, the implications of a contrary holding might be extreme: state courts might be ousted from jurisdiction over interpretation of all compensation clauses in contracts by which the interstate seller obtained inputs necessary to his production.

59 Atlantic Refining Co. v. Public Service Comm'n of New York, 360 U.S. 378, 388 (1959).

60 Sunray Mid-Continent Petroleum Co. v. FPC, 364 U.S. 137, 147 (1960).

61 *See* NGPA § 121.

62 Transcontinental Gas Pipeline Co. v. State Oil & Gas Board, 457 So.2d 1298 (Miss. 1984), *jurisdiction noted,* 53 U.S.L.W. 3867 (U.S. Sup. Ct., 27 Mar. 1985). The case, and the issue of preemption of ratable take orders generally, are discussed at greater length in Williams, *Federal Preemption of State Conservation Laws After the Natural Gas Policy Act: A Preliminary Look,* 56 U. Colo. L. Rev. 521 (1985).

63 457 So.2d at 1331.

to entail invalidation.[64] After all, a targeted price, bearing no particular relation to undistorted market conditions, is inherently involved in setting the production ceiling.

Until recently, however, the resale price of interstate pipelines has been held below market-clearing levels by federal regulation. Consequently, it has been a *federal* price, not a state target, that has determined volume. (As the average price for interstate gas was below market-clearing levels until 1982, the implicit volume was, of course, full-throttle production.) Under current market circumstances, however, with a large percentage of interstate gas deregulated at the wellhead, and with increasing flexibility in gas transportation, federal law may be characterised as inching towards an integrated natural gas market throughout the United States (or even North America), with prices potentially determined by conventional market forces. In such a market, state market demand prorationing of gas, as applied to an interstate pipeline, would seem to conflict with the basic purposes of the regulatory scheme.

Conventional allowables and well-spacing orders, however, seem likely to be valid. It is true that they implicitly involve price questions: the trade-off between the costs of extra wells and the value of accelerating production, for example, could obviously vary depending upon assumptions as to price trends. But, whereas in market demand prorationing the state regulators seem to be *determining* prices (except where federal regulations have done so), it seems probable that, insofar as they consider prices in setting basic allowables and well-spacing rules,[65] they genuinely view them as *given* by the market.

B. Federal Intervention in Crude Oil Markets

We may divide federal intervention in crude oil markets in this century into two major eras. The first, 1913-1973, was marked by a pronounced tilt in favour of producers. It appears to have caused considerable misallocation of resources (as measured by conventional efficiency criteria) and ambiguous national security results. The second, 1973 to the indefinite future, has been marked by a powerful effort to capture economic rents that producers would otherwise have received. It too has had adverse effects on efficiency and almost certainly on national security.

The second era includes a segment, 1973-1981, in which Congress pursued the capture of economic rents by the most awkward of means — price controls. The results paralleled those of the comparable regulatory "mismatch" for

[64] This would be true only if there were a sufficient jurisdictional nexus between the sale and federal regulatory jurisdiction. For a review of the various possible types of jurisdiction, *see* Williams, *supra* note 62.

[65] There is some evidence that the Department of the Interior, in setting allowables for production by federal oil and gas lessees, may seek simply to maximise physical production from the pool subject to the constraint of not letting the extractor's rate of return fall below some acceptable level. *See* S.L. McDonald, The Leasing of Federal Land for Fossil Fuels Production 128-30 (1979).

natural gas:[66] unusually adverse effects on efficiency, a rigid freezing of consumption patterns, and forfeiture of the sort of economic advantages that free trade between the states could provide. The first era was one of stable or declining prices, the second one of generally rising prices.[67] A possible inference is that the national political system is mainly responsive to efforts to protect producers and consumers, as the case may be, from wealth *shifts* (from the *status quo ante*) that might result from the free interplay of market forces.

A virtual precondition of successful redistributive efforts was the nation's containing within its borders petroleum supplies large enough to meet at least a very large portion of domestic demand. As to crude oil, the United States clearly differs in this respect from the EEC.

1. The First Era: 1913-1973

The primary tools of this first period were favourable tax provisions, federal enforcement of market demand prorationing, and quotas limiting the importation of foreign oil.

(a) Tax Advantages

Favourable tax treatment started with a 1913 provision allowing deduction of 5 percent of the gross value of production. Congress in 1926 adopted the approach that prevailed until 1969, entitling the recipient of oil or gas income to deduct 27.5 percent of the gross income (subject to a limit of 50 percent of net income). Changes initiated in 1969 have excluded integrated petroleum producer-refiners from the advantage and have reduced the rate to 15 percent.[68]

The expected effect of such a provision would be to misallocate resources towards the petroleum industry, resulting in greater production and lower prices than would otherwise have prevailed. Holders of petroleum resources (including physical or human capital specialised in such production) would for some time earn above-market profits. Extra capital would, however, be drawn into the industry until the rate of return at the margin was the same as for other activities.[69]

(b) Federal Enforcement of Market Demand Prorationing

In the Connally Hot-Oil Act of 1935[70] Congress gave its blessing, and the advantage of its enforcement powers, to the market demand prorationing (MDP) described above.[71]

[66] *See supra* text accompanying notes 25-35.

[67] *See* S.H. Schurr, J. Darmstadter, H. Perry, W. Ramsay & M. Russel, Energy in America's Future: The Choices Before Us 93-94 (A Study by the Staff of Resources for the Future, 1979).

[68] Internal Revenue Code § 613A.

[69] *See* Mead, *The System of Government Subsidies to the Oil Industry*, 10 Nat. Res. J. 113, 115-16, 123 (1970).

[70] 15 U.S.C. para. 715 *et seq.*; and *see supra* Ch. II, at p. 89. The statute is still on the books.

[71] *See supra* Ch. II, text accompanying notes 255-70.

The expected effect would be to enable oil producers both to correct the "common pool" distortions noted above and to enjoy some of the benefits of cartelisation, with higher prices and lower production levels than competition would produce.

Over the long run, one would not expect the effects to include a higher-than-normal return on investment in the petroleum industry: investors would presumably bid up the prices of petroleum-related capital until return at the margin equalled the return in other industries.

(c) Import Quotas

In 1959 President Eisenhower imposed mandatory quotas on the importation of foreign oil. The quotas were terminated by President Nixon in 1973. Among the results were the protection of United States producers from foreign competition and the requirement that consumers pay more for oil than if the imports had been allowed (a differential estimated at somewhat over $1 per barrel for early 1968, for example, out of a domestic price in East Coast ports of about $3.40).[72] Resources with a value of, say, $3.40 were, therefore, used to obtain domestic oil when its equivalent was available from the Middle East at perhaps $2.20.

National security implications of the quotas are ambiguous. On the one hand, while they were in effect they caused the country to consume more of its own oil resources so that in a physical sense less domestic oil was available in the ground at times of crisis (*e.g.*, 1941-1945 and 1973) than would otherwise have been the case. On the other hand, the price maintenance encouraged exploration for oil, and sustained a larger domestic petroleum industry than would otherwise have been in existence at the various crisis periods.

The three types of federal intervention — favourable tax treatment, enforcement of MDP (insofar as it was used for purposes other than merely to remedy the "common pool"), and import quotas — all distort efficiency, although in opposite directions (the first tending to increase supply and lower price, the second to decrease supply and raise price, and the third to increase domestic supply and price). Ascertaining the net impact would be a complex and probably impossible chore. All three have in common, however, a tendency to provide immediate wealth benefits for domestic oil producers.

The harmonious mood associated with these early interventions is striking to an observer more directly familiar with the 1970s and early 1980s. If the Connally Hot-Oil Act, for example, represented a victory for the oil-producing states, it was a victory without a battle. Predecessor legislation, included as section 9(c) of the National Industrial Recovery Act of 1933, had been struck

[72] *See* Mead, *supra* note 69, at 113-14, 120-22; J.P. Kalt, *supra* note 40, at 6-8. Kalt cites estimates of Bohi and Russell that the quotas increased domestic producers' rents by $2.3 billion in 1969, transferred $0.8 billion to refiners holding import quotas, cost consumers $5.4 billion in consumer surplus, and caused a deadweight social loss of $2.3 billion.

down by the United States Supreme Court on 7 January 1935, on the ground that, in authorising the President to impose the restrictions, it had made too broad a delegation of Congressional authority.[73] By February 25, both houses of Congress had passed and the President had signed the Connally Act itself.[74] The votes were all voice votes,[75] and the most substantive objection was based on seemingly principled federalist grounds — concern that the Act was an excessive exercise of the federal government's power.[76] Representatives of the oil-importing states were evidently oblivious of any risk that the legislation might assist producing states in effecting a wealth transfer. The early New Deal's delight in government-managed cartelisation seems to have swept aside any doubts.

2. The Second Era: 1973 to Indefinite Future

The increase in the world petroleum price that started in 1973 clearly represented an opportunity for the then holders of American petroleum assets (both reserves and extractive physical and human capital) to enjoy dramatic upward shifts in the value of those resources. If all oil were priced at the cost of the marginal supply,[77] which in 1974 was OPEC oil priced at about $12 a barrel, owners of all lower-cost oil would receive very large economic rents. Congress moved to capture those rents and reallocate them, initially by a programme of price controls and "entitlements", later by a "Windfall Profit Tax".

(a) 1973-1980: Price Controls and Entitlements (EPAA/EPCA)

Although direct federal price control of crude oil originated in August 1971 with President Nixon's freeze of the prices of virtually all goods and services, they developed their own special rent-capturing character in August 1973, when the Cost of Living Council developed a two-tier price control programme, exempting "new oil" and controlling only "old oil".[78] Price controls on other commodities lapsed gradually, partly by Presidential order and partly

[73] Panama Refining Co. v. Ryan, 293 U.S. 388 (1935).

[74] The Connally Act differed from § 9(c) in that: (a) Congress expressly imposed the prohibition, with the President's discretion having the *form* of a power to render the prohibition inoperative; and (b) Congress gave the President a vague set of instructions to guide him in the exercise of that discretion.

[75] *See* 79 Cong. Rec. S764, H2150 (22 Jan.1935 and 18 Feb. 1935) for the original votes, and 79 Cong. Rec. S2513, H2502 (22 Feb.1935) for approvals of the House-Senate conference report.

[76] Senator Borah of Idaho objected that in the past Congress had barred the interstate transportation of goods only when the goods themselves had some evil characteristic. 79 Cong. Rec. S763 (22 Jan. 1935).

[77] As under market conditions it would be.

[78] *See* 38 Fed. Reg. 22,538 (1973).

by the expiration in 1974 of the President's authority under the Economic Stabilization Act.[79]

The programme of multi-tier control over crude oil prices continued, however, under the authority of the Emergency Petroleum Allocation Act of 1973 ("EPAA")[80] and the Energy Policy and Conservation Act of 1975 ("EPCA").[81] They ended in January 1981 with President Reagan's exercise of his decontrol authority under EPCA.[82] Although EPCA modified EPAA, the basic principles were the same throughout, and we shall refer to them collectively as the "EPAA/EPCA" controls.

(i) The Mechanics of the EPAA/EPCA Regime

The guiding principle was to define vintages of oil, and to establish for each vintage a price ceiling loosely related to its cost of production (or to exempt it from controls altogether). Thus, "new oil" was distinguished from "old oil" (the precise definitions varying over time) and either exempt (under the Cost of Living Council regulations and EPAA) or subject to a higher ceiling (under EPCA). For most of the period "stripper oil" (oil from wells producing less than 10 barrels per day and therefore marginal) was exempt.[83] "Incremental tertiary oil" (oil produced by tertiary recovery methods in excess of certain base levels) was exempted in 1978.[84] In fact, of course, units of oil in any of the categories will vary in cost, so the controls necessarily rendered some oil in non-exempt categories submarginal. But the difference between market price and each ceiling price could be seen as a rough approximation of the economic rent.

As had been true of natural gas, demand for crude oil in lower tiers obviously exceeded the supply. Prior to the autumn of 1974, this was handled by freezing refiners' rights to lower-tier oil in their historical (1972) relationships to their crude oil suppliers, so that each refiner was entitled to the same fraction of any crude producer's current output as he had been purchasing from that supplier in the base period.[85]

[79] *See* Economic Stabilization Act of 1970, P.L. No. 92-210, 85 Stat. 743, *as amended by* P.L. No. 93-28, 87 Stat. 27.

[80] P.L. No. 93-159, 87 Stat. 627.

[81] P.L. No. 94-163, 89 Stat. 871.

[82] *See* Executive Order 12,287, 46 Fed. Reg. 9909 (1981).

[83] Congressional vacillation on "stripper" oil was extraordinary. It initially exempted stripper oil by a provision of the Trans-Alaska Pipeline Authorization Act of 1973, para. 406, P.L. No. 93-153, 87 Stat. 576, then extended the exemption in EPAA, then nullified the exemption in EPCA, but finally recreated it in the Energy Conservation and Production Act of 1976, para. 121, P.L. No. 94-385, 90 Stat. 1125.

[84] This occurred by virtue of the Economic Regulatory Administration's adopting 10 C.F.R. para. 212.78, 43 Fed. Reg. 33689 (1978), pursuant to authority granted by the Energy Conservation and Production Act, *supra* note 83, at para. 122.

[85] *See* J.P. Kalt, *supra* note 40, at 12-13.

Quite apart from its other drawbacks, the buyer-seller freeze gave a competitive advantage to refiners with large purchases in the base period from a crude oil producer with a large current fraction of lower-tier crude. Regulations promulgated in the autumn of 1974 addressed that problem by establishing an entitlements programme that survived until the end of controls. Each refiner received, for any month, entitlements to controlled crude oil equal to the number of barrels of controlled crude oil that the refiner would have used in the previous month if it had operated using the national average ratio of controlled to uncontrolled crude oil. A refiner that in any month purchased *more* controlled oil than the amount for which he had received entitlements was required to purchase entitlements to make up the difference. A refiner purchasing *less* controlled oil than his issued entitlements could sell the surplus entitlements.

The effect of the entitlements was that each refiner purchased all its oil at a price equal to the weighted national average price for controlled and uncontrolled oil. This would obviously be so for a refiner buying oil with the national average ratio of controlled to uncontrolled oil, *i.e.*, neither buying nor selling entitlements. For others, it came about because the purchase price of an entitlement was set at the difference between refiner cost of controlled and refiner cost of uncontrolled oil. Thus, for example, a refiner that bought no controlled oil in a given month would be able to sell all its entitlements to such oil, so that for the fraction of its purchases corresponding to the national fraction of controlled oil, it would get the benefit of the controlled price.[86] (For example, suppose that in a given month controlled and uncontrolled oil each constituted half of United States consumption and that refiner A used a total of 100 barrels, all uncontrolled. If controlled oil sold for $ 6 a barrel and uncontrolled for $ 12 a barrel, refiner A would be able to sell entitlements to 50 barrels of controlled oil at $ 6 each, so that it would effectively pay $ 12 a barrel for 50 barrels and $ 6 a barrel for the other 50.)

In economic effect, the controls-and-entitlement programme was equivalent to (1) allowing domestic producers to sell all oil at the weighted average of controlled and uncontrolled oil, (2) taxing producers of controlled oil by the difference between that average price and the controlled price, and then (3) using the tax proceeds to issue vouchers to refiners to buy uncontrolled oil.[87] As imported oil was as eligible for the entitlements subsidy as any other uncontrolled oil, a prime effect of the subsidy was to expand American consumption of imported oil.

EPAA/EPCA also provided for price controls on refined petroleum products. This was accompanied by an allocation programme, primarily based on historical purchase levels.[88] As in the case of natural gas, this freezing of relative consumption patterns was inconsistent with the efficiency requirement that the value of each user's marginal unit be the same. Consumption patterns

[86] *See id.* at 13-14.
[87] *See id.* at 53-54, 65-66.
[88] *See* 10 C.F.R. para. 211.9.

Table 9
Estimated Distributional Effects of Controls and Entitlements (Most Likely Case): 1975–1980 (Billions of 1980 Dollars)

	Crude oil producers	Petroleum refiners	Petroleum product consumers	Deadweight social loss
1975	–23.9	+15.0	+ 6.9	–2.0
1976	–18.9	+10.2	+ 6.8	–1.9
1977	–18.7	+10.4	+ 6.4	–1.9
1978	–14.3	+ 8.5	+ 4.7	–1.1
1979	–32.6	+21.8	+ 8.3	–2.5
1980[a]	–49.6	+31.7	+12.2	–5.7

SOURCE: Based on data from MONTHLY ENERGY REVIEW (US Department of Energy, Energy Information Administration, 1974–).

[a] Annual rate. The row for 1979 in essence states Kalt's estimate that in that year the programme transferred $ 8.3 billion to petroleum consumers from crude producers, but at a cost of $ 32.6 billion to those producers. The difference, by his estimates, went in part to refiners and in part constituted deadweight social loss. These effects reach a peak in the first deadweight social loss. These effects reach a peak in the first quarter of 1980, with $ 49.6 billion captured from producers, $ 12.2 billion of consumer gains, $ 31.7 billion transferred to refiners, and $ 5.7 billion of deadweight social loss (all in annual rates). The figures for 1980 are highest because they represent the time of the largest gap between world market prices and the US controlled price.

changed in response to the aggregate shortage, but supplies did not shift in response. Gasoline shortages, for example, were acute in metropolitan regions, where drivers remained in response to price increases and in fear of being caught short in outlying areas; in the latter, by contrast, supplies were generally ample. Under market conditions, metropolitan gasoline dealers would have bid the supplies away from the others; the allocation programme forbade them to do so.

One effort to provide flexibility, interestingly, employed the states as units of reallocation. Under "state set-aside" provisions, 5 percent of the total gasoline allocation to each state was withdrawn from allotment under the basic pro rata distribution programme. State energy agencies were authorised to employ this 5 percent to mitigate the effects of the Procrustean pro rata allocation.[89] This is obviously a far cry from the vision of the United States as a single market. Supply patterns were frozen in terms of very small units — the individual retailers — and the modest 5 percent margin of flexibility was confined within each state.

Rent distribution by these devices does not come cheap. In Table 9 we reproduce Joseph Kalt's estimates of losses to crude oil producers, gains to re-

[89] *See* 10 C.F.R. para. 211.17.

finers and consumers, and the deadweight social loss.[90] Kalt estimates that the programme tended to increase United States crude oil imports substantially, ranging from about 580 million barrels in 1975 (or about 40 percent of total crude oil imports), with a cost of about $11.5 billion, to an annual rate of about 900 million barrels in the first quarter of 1980 (or about 45 percent of total imports), with a cost of about $33 billion.[91]

[90] *See* J.P. KALT, *supra* note 40, at 216. We have supplemented his table to include a column expressly devoted to deadweight social loss; this can be calculated directly from his table at 216, but also appears in tables at 191 and 201. He calculates deadweight social loss from the entitlement programme as (1) the true cost of the additional barrels of petroleum imported as a result of the programme, minus (2) the true market value of such additional barrels. The basic point is that the subsidy to the purchase of imported crude causes refiners to push their production out further along the consumers' demand curve than they would under market conditions. The real cost of the additional barrels is the world price (times the quantity), but their value to consumers is only the area under the demand curve. In Kalt's graph, reproduced below, the deadweight social loss is the triangle XYZ, calculated as C*C'ZY minus C*C'ZX.

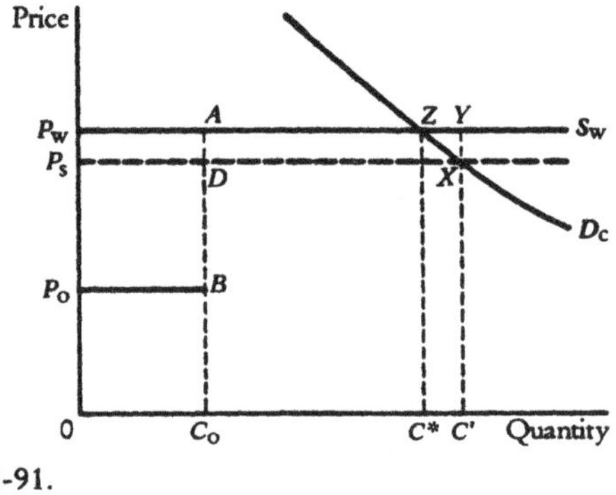

See id. at 49, 188-91.

Deadweight social loss from the price controls is calculated as (1) the value of oil production foregone as a result of them, as measured by consumer willingness to pay, minus (2) the value of such production as measured by the cost of production. Among possible sources of error are the following: (1) Each imported barrel of crude oil imposes an external cost in the form of national security risk. Disregard of this cost causes the estimation to understate the loss. (2) The figures omit government enforcement cost and compliance costs in the private sector (*e.g.*, costs of legal advice as to the meaning of the requirements), and thus understate the loss. (3) Each barrel of US production kept in the ground is available for later production; to the extent that computation of the cost of barrels not produced as a result of the price controls disregards this feature (namely, user cost, *see supra* notes 7 and 8), the loss is overstated.

[91] *See id.* at 191 & 201. Percentages of crude oil imports are calculated by comparison with import figures in DEPARTMENT OF ENERGY/INTERNATIONAL AFFAIRS OFFICE OF MARKET ANALYSIS, INTERNATIONAL ENERGY INDICATORS 18-19 (DOE/IA-0010/14, Aug.-Sept. 1981).

In the Congressional adoption of EPAA/EPCA, the rhetoric of wealth transfer — a desire to capture "obscene profits" — was dominant. Proponents occasionally, however, pressed one argument relating directly to efficiency, the "full employment" argument. Simply put, the argument runs that in the absence of price controls the abrupt increase in oil prices will shift purchasing power from consumers to producers (both domestic and foreign), that consumers will respond by reducing their purchases of commodities other than oil (a reduction that will not be immediately offset by equal increases in purchases by oil producers), that aggregate demand will therefore fall, and that unemployment of labour and other productive resources will result.[92]
impact of the increased scarcity evidenced by the price change: comparing the period after 1973 with the 1960s, each barrel of oil consumed by the nation cost it more in real resources, either the resources employed to produce the oil domestically (*i.e.*, price less economic rent) or the resources that had to be transferred to foreign suppliers in exchange for their oil. No one could seriously argue that a controls-and-entitlements policy would offset this impact.[93]

The full employment argument has a number of vulnerabilities that make us think that its role must have been slight. First, it seems likely that a shift of purchasing power to an OPEC country would have a more immediate adverse effect on aggregate United States demand than a shift of equivalent purchasing power to American oil owners and explorers. A serious proponent of the theory would surely want to consider whether there would be any net full employment benefits in a policy which, like controls-and-entitlements, (a) reduced the aggregate shift of purchasing power away from American consumers, but (b), as to the amount of purchasing power which *was* allowed to shift to producers, increased the proportion going to foreign ones.

Second, even in the neo-Keynesian view of macro-economic policy, inadequate information plays some role in obstructing the adjustment to a macroeconomic shock; the "stickiness" of wages and prices is not wholly divorced from the economic factors' failure to grasp the implications of the shock.[94] (In other models, information gaps are much more critical.) Accordingly, it would be paradoxical, to say the least, if a policy that concealed critical information about the shock, namely, its impact on relative prices, were to facilitate adjustment to the shock.[95]

[92] *See* H.R. 340, 94th Cong., 1st Sess. (1975), to accompany H.R. 7014, 94th Cong., 1st Sess., at 8, 144-48 (1975), *reprinted in* 1975 U.S. CODE CONG. & AD. NEWS 1770, 1885-89. For a more sophisticated exposition, *see* FORD FOUNDATION STUDY GROUP, ENERGY: THE NEXT TWENTY YEARS 157-64 (1979).

[93] *See* FORD FOUNDATION STUDY GROUP, *supra* note 92, at 155-57.

[94] *See* A.M. OKUN, PRICES AND QUANTITIES: A MACRO-ECONOMIC ANALYSIS 37-41 *et passim* (1981); Solow, *Arthur Okun's Last Work (Book Review)*, 65 PUBLIC INTEREST 91, 94-99 (Fall 1981).

[95] *See* J.P. KALT, *supra* note 40, at 224-25.

Third, a windfall-profit-tax approach could achieve any full employment benefits that controls-and-entitlements could obtain, with considerably reduced adverse effects on efficiency apart from full employment problems. (Windfall profit tax revenues could be distributed to consumers in the form of transfers or cuts in other taxes. But by allowing the price to float to world market levels, it would inform energy users of the real resource cost of the marginal unit of oil and induce them to make such substitutions as were efficient at that price.) Accordingly, a person for whom the full employment efficiency argument was uppermost could be expected to prefer a windfall profit tax to controls and entitlements.[96]

(ii) Appraisal of the EPAA/EPCA Policy

The results of the policy seem negative, or likely to be negative, by any criterion except that of wealth redistribution. At a time of professed anxiety about dependence upon foreign suppliers,[97] it diminished United States supply and increased demand. On the premise that the national security costs of dependence upon foreign oil were likely to grow beyond the 1970's levels, one might argue that by depressing current American production, the policy kept more oil in the ground, the better to handle more acute national security problems later on. Since the remainder of the century seems almost sure to manifest a gradual substitution away from liquid petroleum, however, the premise seems dubious.

In efficiency terms, the policy induced Americans to consume petroleum products whose real value to them was less than the world market price — the real price paid for the marginal barrel — but higher than the United States price resulting from control. It led Americans to forego the production of oil at prices between the various ceilings and the world market price. The artificial

[96] G. Horwich & D.L. Weimer, Oil Price Shocks, Market Response, and Contingency Planning (1984), examine the evidence as to the scale of these shifts in purchasing power, which they denominate "oil price drag". They conclude that the "drag" probably amounted to 1.86% of GNP in 1974 (year of highest percentage) and 1.39% in 1979 (year of the highest dollar amount). *Id.* at 161. They argue that even these figures overestimate its significance, contending that, since two thirds of oil purchases are for oil as an intermediate product, the appropriate reference is total transactions, which are at least double the size of GNP.

Whatever the scope of oil price drag, Horwich and Weimer argue that it can best be addressed with a policy of monetary ease. *Id.* at 165-66.

Horwich and Weimer also consider whether price controls on crude oil would diminish or increase inflation. They believe the latter. The price controls would cause micro-economic inefficiencies, *i.e.*, a lower real value of output. Fewer goods being chased by the same number of dollars means greater inflation. At least assuming identical levels of money supply and velocity, the conclusion seems sound. *See id.* at 95.

[97] *See, e.g.*, H.R. 340, *supra* note 92, at 3, *reprinted in* 1975 U.S. Code Cong. & Ad. News 1765.

shortages generated by control of refined petroleum product prices led to mandatory allocation that froze consumption patterns and negated any chance that the marginal value in use for every consumer across the country would be equal. Chances of the policy's contributing to full employment seem dubious.

One might advance an efficiency argument in favour of EPAA/EPCA paralleling the national security argument raised above (and rejected): that inability of future consumers of oil to bid for its use constitutes a market failure and that price controls, tending to keep the oil in the ground, offset such a market failure. This formulation of the alleged market failure is clearly too broad, for loss of the increment in value to be derived from withholding reserves for future extraction and sale is clearly one of the opportunity costs of present extraction, at least if unitisation is freely enough available to remedy the common pool problem.[98] As owners of reserves incur this opportunity cost when they engage in present extraction, they have a substantial pecuniary incentive to preserve resources for future consumers. It would, therefore, be appropriate to reformulate the argument, perhaps in terms of an argument that the interest rate at which owners discqunt future returns, in order to compare them with present ones, is either inefficiently or unethically high, so that government intervention slowing the rate of present consumption is desirable.[99]

However a hypothetical Congress *might* have addressed such concerns, they appear to have played no material role in the thinking of the Congresses that adopted EPAA/EPCA. Indeed, the proponents of the legislation expressed their desire not to diminish production; they answered arguments that the legislation would do so with conclusory assertions that supply was unresponsive to price.[100] In EPCA itself, they took steps to pressure federal oil lessees to produce *faster*, by transforming a production *ceiling* concept, the "maximum efficient rate" or "MER", derived from state conservation efforts, into a *floor* on those lessees' production levels.[101] If EPAA/EPCA had any tendency to make the intertemporal allocation of oil more efficient, that effect was entirely coincidence.

[98] *See supra* Ch.II, at pp. 86-87

[99] Some of the arguments are reviewed in Williams, *Running Out: The Problem of Exhaustible Resources*, 7 J. LEGAL STUD. 165 (1978). Of course this justification would still fail to explain a policy that exposes consumers to a price lower than the actual value paid to foreign suppliers. The miscue to consumers will lead them to invest in capital (*e.g.*, large cars, poorly insulated houses) that is inappropriate for the present or future real price of oil.

[100] *See, e.g.*, H.R. 340, *supra* note 92, at 39, *reprinted in*1975 U.S. CODE CONG. & AD. NEWS 1801.

[101] *See* Energy Policy and Conservation Act of 1975, para. 106, 42 U.S.C. para. 6214 (1975). The Federal Coal Leasing Amendments Act of 1975 [actually adopted 4 Aug. 1976], P.L. No. 94-377, 90 Stat. 1083, reflects a similar desire to prod current production up to levels in excess of what coal lessees would select in response to market conditions.

A recent study of Senate voting patterns on crude oil issues in 1973-1977 indicates that regional interests played an important role. Kalt, using indicators of the importance in each state's economy of crude oil[102] and of energy consumption,[103] found both factors to have considerable explanatory power for a series of votes affecting distribution of wealth as between crude oil producers and energy consumers. Ideology, measured by the Senator's ADA voting record, played a significant role, but by no means the overwhelming one which it did in the Mitchell studies of natural gas deregulation.[104]

(b) 1980 to the Indefinite Future

Passage of the Windfall Profit Tax Act[105] in April 1980 (retroactive to March 1980) introduced the second segment of the post-1973 era. Substitution of the tax for controls and entitlements modified the trade-off between redistributive purposes and concern over efficiency, with effects much less disastrous to efficiency. Decontrol allowed prices to move to world levels, so that American consumers had an incentive to use oil only when the cost to them of foregoing that oil (either in lost pleasure, or the cost of some substitute such as insulation) was higher than the oil's cost in real resources. Non-price allocation, through administrative fiat or queueing in gasoline lines, became unnecessary.

[102] The variable is measured by the product of (1) state crude oil production in 1975 and (2) the difference between (a) the average landed cost of foreign oil and (b) the state's average price of crude oil in 1975, expressed as a fraction of state personal income. *See* J.P. KALT, *supra* note 40, at 257.

[103] The variable is measured by total state expenditures on energy in 1974, expressed as a fraction of state personal income. *See id.*

[104] Kalt's key equation, in which PROCRUDE represents votes favourable to crude oil producers and PROADA signifies a senator's voting record as valued by the Americans for Democratic Action (ADA), is as follows (*t*-statistics in parentheses):

$$\begin{aligned} \text{PROCRUDE} = \; & \underset{(0.39)}{0.085} + \underset{(5.91)}{3.596}\ \text{CRUDE} - \underset{(-0.83)}{0.786}\ \text{REFINE} \\ & -\underset{(-0.04)}{0.076}\ \text{SMALL REF} - \underset{(-3.35)}{9.416}\ \text{ENERGY USE} \\ & -\underset{(-19.07)}{0.51}\ \text{PROADA} \end{aligned}$$

$R^2 = 0.52$; F-statistic $= 20.56$

Id. at 265-66.

Because there is no single unit in which to measure ideology and the roles of crude oil production or energy consumption in a state's economy, the above equation does not permit one to say that one factor was more important than the others. It does, however, support the conclusion that all three factors were distinctly important.

It is ironic that the variables for the importance of refiners and small refiners in a state's economy lacked statistical significance (*id.* at 266), since refiners appear to have been the big winners from the EPAA/EPCA policy.

[105] P.L. No. 96-223, 94 Stat. 229.

Adverse production effects remain. Again the principle of multiple tiers is employed; the amount subject to tax is, for each barrel, the difference between its actual selling price and a statutorily determined "base price" for the tier to which it belongs. The nation foregoes production of some of the oil, in any tier, which could be produced at a cost between its "base price" and the world market price. A Congressional Budget Office study, prepared while the tax statute was progressing through Congress, estimated that the Senate Finance Committee bill (considerably milder than the final product) would reduce United States production at an annual rate of about 70 million barrels in 1985 and an annual rate of about 105 million barrels in 1990.[106] A 1980 Department of Energy study of the tax actually enacted estimated continuation of the tax would generate extra petroleum imports amounting to a half quadrillion BTU in 1990, or about 90 million barrels.[107]

The Windfall Profit Tax is a rare exception to the general rule that, although taxes seem far superior to price controls as a device for rent redistribution, legislatures prefer price controls. What accounts for this gulf between public welfare and the political process? Part of the explanation may lie in the difficulty of identifying the beneficiaries of a tax. Because dollars are fungible, it is almost impossible to establish a causal link between the tax and any specific spending programme, to say that any particular tax made any particular spending programme possible. Even "earmarking" specific tax revenues for specific purposes, as did the Windfall Profit Tax Act (earmarking them for energy aid to the poor and for synthetic fuel subsidies),[108] fails to establish the causal relationship. First, nothing in the earmarking can make it binding upon future Congresses. Second, and more fundamental, the tax revenues could only be

[106] CONGRESSIONAL BUDGET OFFICE, THE WINDFALL PROFIT TAX: A COMPARATIVE ANALYSIS OF TWO BILLS 15 (Nov. 1979).

[107] *See* DEPARTMENT OF ENERGY/ENERGY INFORMATION ADMINISTRATION, ENERGY PROGRAMS/ENERGY MARKETS: OVERVIEW 59 (DOE/EIA-0201/16, July 1980).

Changes in the price of oil since these studies were made appear likely to drastically reduce both the tax's total take and its negative effect on production. One may calculate the annual average revenue yield expected by the tax's sponsors by dividing the aggregate yield that was to trigger the start of a three-year-phase-out period (*i.e.*, $227 billion) by 10 (*i.e.*, the number of years between the tax's taking effect (1980) and the alternate date that would trigger the start of phase-out (1990)). In sharp contrast to that expected average annual yield of nearly $23 billion, the current estimate of annual net yield is of less than $6 billion for calendar year 1983, gradually declining to less than $3 billion in 1989. *See* OFFICE OF THE SECRETARY OF THE TREASURY, OFFICE OF TAX ANALYSIS, WINDFALL PROFIT TAX LIABILITIES AND RECEIPTS (25 Jan. 1984, Documentation of 1985 Fiscal Year Budget).

The yield drops more rapidly than the price of oil because the tax is highly leveraged, taking a very high fraction (in some cases as much as 70%) of the portion of a barrel's gross revenue *above* a statutorily prescribed "base price", but taking no part at all of amounts below that price.

[108] *See* Crude Oil Windfall Profit Tax Act of 1980, § 102, P.L. No. 96-223, 94 Stat. 255 (codified at 31 U.S.C. para. 553(b)(3) (Supp. V 1981)).

said to "cause" the expenditures for which they were earmarked if one knew for a certainty that Congress would not have ordered equivalent spending in the absence of the revenues; but, particularly in an era when deficit spending is the norm, there is no way to prove such a thing. Thus proponents of a rent-redistributing tax find it difficult to secure the popular support that could derive from linking it to popular spending programmes.[109]

II. The European Communities

Within the Community framework the conditions under which central intervention could impede the free flow of energy across its component parts, whether with redistributive or other ends in view, have hardly yet arisen. This is because there has been very little effective central intervention in the Community energy market at all: indeed it is still not possible to speak of a single Community energy market. In this section we look at the supply policy provisions contained in the three Treaties, and at the experience to date of their operation (or non-operation). This should permit us to assess, by reference to the American position, the level of development of centralised energy policy in the Communities and the likelihood of its avoiding the pitfalls of market fragmentation identified in the United States.

A first obstacle in the way of a single Community market in energy, which we have already discussed in a little detail, is the division of energy matters among three separate treaties. Such a division makes it difficult (though not wholly impossible) to translate to the Community level the decisions about the short- and long-term place of specific energies in the supply profile which have been the central feature of national energy debates since the 1973 oil crisis, and which were not without significance even in earlier, more relaxed times. The difficulty is compounded by the fact that the three Treaties apply different principles, and offer different machinery, for the resolution of questions of energy supply policy. These differences were justified by what were seen to be different kinds of problems presented to Europe by the supply of the various energy-related natural resources: coal, oil, and nuclear materials. In each case, however, either the problem was misperceived, or the machinery proved inadequate, or the will to operate it, among Commission, Council and Member States, was lacking.

A. The Euratom Regime

The Framers of the Euratom Treaty designed a fully centralised system of supply of nuclear materials. This is the system whereunder a Community

[109] Senator Long, floor manager of the bill in the Senate, seems explicitly or implicitly to have recognised both these vulnerabilities of earmarking. *See* 126 CONG. REC. S3042 (26 Mar. 1980).

agency, the Euratom Supply Agency (ESA), enjoys a virtual monopoly within the Community of importation, purchase, and supply of ores, source materials, and special fissile materials,[110] among which the most important are uranium ore, unenriched uranium (which is defined as a source material), and plutonium and enriched uranium (which are defined as special fissile materials). The system makes different provision for special fissile materials on the one hand, and ores and source materials on the other. Largely in order that the Community may exercise adequate supervision over their use and penalise misuse with adequate sanctions, including, ultimately, the withdrawal of such materials from the user, the property in special fissile materials is vested in the Community.[111] This rule applies both to imports and domestic production, but only if the materials are subject to the Community's safeguards system,[112] a qualification which excludes from Community ownership materials intended to meet defence requirements.[113] Community ownership does not preclude the use and consumption of special fissile materials by their possessors or connected undertakings, unless the ESA exercises its right of option to use and consume the materials.[114]

While the Community's safeguard system also applies in relation to ores and source materials, the need for central control of the materials themselves was thought to arise principally in connection with the securing of equal access to such materials for users in all Member States. (This equal access objective also applies, of course, in relation to special fissile materials, but subject to major limitations on the ESA's exclusive rights in favour of those who have manufactured them.[115]) The instrument for securing this objective was the ESA, a Community organ under the supervision of the Commission, entrusted with the task of matching users' needs with producers' capabilities and with the availability of supplies from outside the Community, and concluding contracts for the purpose of furnishing such supplies.[116] Where supplies fell short of demand, the ESA was to share out supplies proportionately among the relevant orders.[117] To enable it to discharge these functions, the Agency, as already noted, enjoys an exclusive right of importation into the Community,[118] and producers within the Community are bound to offer their production to it.[119]

[110] These terms are defined in Euratom Treaty art. 197.
[111] *Id.* art. 86.
[112] *Id.* ch. VII (arts. 77-85).
[113] *See id.* art. 84, para. 3.
[114] *Id.* art. 62.
[115] *Id.* art. 62(2).
[116] *See generally id.* ch. VI (arts. 52-76).
[117] *Id.* art. 60, para. 5. It is not clear whether the Treaty Framers considered the problems, such as deliberate over-ordering, which could arise with the implementation of this system. It is most unlikely that the system will ever operate in future: *see infra* text accompanying notes 129-41.
[118] *Id.* art. 64.
[119] *Id.* art. 57.

In addition to these centralised supply arrangements, and the common market in nuclear materials already mentioned,[120] the Euratom regime transfers competence over pricing to the Community, providing that prices should normally be determined by the balancing of supply against demand, with which national regulations must not interfere, but that they may also be fixed by the Council. Pricing practices designed to secure a privileged position for certain users in violation of the principle of equal access are prohibited.[121] The Treaty also provides for a common supply policy,[122] involving the encouragement of domestic production, the denial of equal access to other Community sources to Member States that do not exploit their own resources, commercial stockpiling by ESA and emergency stockpiling by the Commission.[123]

The centralisation of essential power over natural resources in Euratom is in striking contrast to the position obtaining in the other Communities where, in the absence of emergency situations, Community objectives are to be achieved essentially by the prohibition of market-compartmentalising behaviour by Member States. The contrast is unreal, because the supply arrangements for nuclear fuel in Europe have never operated in accordance with the Treaty text, and are unlikely ever to do so in the future. The non-application of the Treaty may be linked to a general weakness of Euratom stemming from an early failure of the will of the Member States to co-operate effectively in the nuclear sector, but it has more specific roots in the development of the supply situation. In the first place, the feared dearth of supplies of source materials, which was a principal cause of the specific design of the Treaty, did not materialise. By 1960 the Council had adopted a decision restricting the Agency's exercise of the right of option and instituting simplified procedures regarding supply contracts which effectively left the parties to conclude their own contracts, subject to a right of supervision of the Agency which has been of little practical significance.[124] This situation still continues. Second, the Treaty has never had any practical effect on the disposition of the Community's own uranium supplies. Workable supplies have hitherto been found only in France and have never been effectively subjected to the Treaty regime, being employed partly for French defence purposes,[125] partly for the fuelling of French nuclear power stations, which enjoyed an initial derogation under article 223 and which have since benefitted from the "connected undertakings" provision of article 62. Third and, in the opinion of one commentator, most

[120] *See supra* Ch.II, text accompanying notes 5-10.
[121] *See* Euratom Treaty arts. 67-69.
[122] *Id.* arts. 70-72.
[123] Neither form of stockpiling has ever in fact taken place.
[124] *See* EAEC Commission Decision of 5 May 1960, 1960 J.O. 776, 1959-1960 O.J. (spec. Eng. ed.) 45; and the Rules of the ESA of 5 May 1960, *id.* at 777 & 46 respectively. Minor modifications were made in 1975: *see* O.J. L 193, 25 July 1975, p. 37.
[125] And thus exempted from the regime of the Treaty under art. 84, para. 3.

important,[126] Member States have not observed the spirit, and perhaps not even the letter, of the provisions which sought to complete the Community's effective control of supplies by substituting the Community for Member States as a party to any agreement for nuclear co-operation negotiated with third states.[127] In the 1960s and early 1970s the Commission appears to have given up the attempt to secure respect for this provision.[128] Some external supplies, particularly of enriched uranium, thus continue to be made under state-to-state-agreements, though the Community has recently taken a more active line in this area.[129]

The Framers of the Treaty appreciated that the supply situation might not develop as expected, and accordingly provided in article 76 that after seven years of operation, the supply chapter of the Treaty should be confirmed by the Council, or new provisions adopted. The Council, however, found it impossible to agree on this question, with the result that, quite contrary to the intention of the Treaty, Chapter VI has never been either confirmed or revised. On this basis, the French Government, accused by the Commission in 1971, before the European Court of Justice, of numerous breaches of the supply provisions of the Treaty, claimed that the activities of the ESA under the supply provisions had become "unreal" and without legal foundation.[130] The rejection of this view by the Court did not hasten the process of reaching a decision on the Chapter. In December 1982, however, the Commission sent to the Council a proposal for revision of the Chapter.[131] Under this proposal, while it

[126] *See* J.G. POLACH, EURATOM: ITS BACKGROUND, ISSUES, ECONOMIC IMPLICATIONS 122 (1964).

[127] Euratom Treaty art. 106.

[128] EURATOM COMMISSION, FOURTH GENERAL REPORT paras. 148, 163 (1961).

[129] Examples of state-to-state agreements include the 1975 Federal Republic of Germany/USSR agreement/contract for uranium enrichment; the France/USSR Agreement for enriched uranium; the 1981 France/Canada/Australia Uranium Agreement; the France/India Agreement of 27 Nov. 1982; and the United Kingdom/Australia Bilateral Agreement, signed July 1979. Article 103, which provides a machinery for checking the compatibility of Member State agreements within the Treaty, has been invoked by the commission in Ruling 1/78, delivered pursuant to EAEC Treaty art. 103(3) (Draft Convention of the International Atomic Energy Agency on the Physical Protection of Nuclear Materials, Facilities and Transports), [1978] E.C.R. 2151. The agreement between the UK and Australia of July 1979, contains a provisional clause to the effect that the agreement exists only until there is an agreement between Euratom and Australia.

[130] Case 7/71, Commission v. French Republic, [1971] E.C.R. 1003, 1013-14.

[131] Proposal for a Council Decision adopting new provisions relating to Chapter VI (Supplies) of the Treaty establishing the European Atomic Energy Community, O.J. C 330, 16 Dec. 1982, p.4; *see* commentary by Allen, *The Euratom Treaty, Chapter VI: New Hope or False Dawn?*, 20 C.M.L. REV. 473 (1983). The proposal has since been amended in minor respects (*see* Doc. COM(84) 606 final/2 (30 Nov. 1984)), but still awaits formal consideration in Council. The amended version of the draft revised Chapter first presented in an Annex to the 1982 proposal, is now to be found at the end of Doc. COM(84) 606 final/2 and is hereinafter cited as *Annex*.

continues to be the Community's objective to ensure that all users receive a regular and equitable supply of source materials, the ESA loses its central position and its contractual monopoly. The preferred means are now threefold. The first is the "unity of the market",[132] to be secured by the prohibition of restrictions on transfers of materials within the Community and on imports, and of conditions governing use and storage within the Community. It will be for the Council to lay down such restrictions and conditions, by way of a Regulation. The second is the retention of the Community's position of primacy in regard to the conclusion of international agreements.[133] Individual Member States may only conclude supply agreements with third states, with the Commission's authorisation, where no appropriate Community agreement exists. Third, there are "solidarity measures",[134] in the form of provision of information by undertakings, Commission aids for uranium prospecting, stockpiling, and special measures to be taken (by the Council) in the event of a significant imbalance between supply and demand. The Agency's role is reduced to one of receiving notification of uranium supply contracts, collecting and disseminating market information, and performing certain other specific tasks as requested by the Commission.[135] Although, on past form, it is unlikely that this proposal will be adopted by the Council in its present form even after the Commission's amendments, it is worth spending a moment to elucidate the current thinking of the Commission on the appropriate framework for nuclear supply policy.

Despite the dismantling of the (hitherto inoperative) provisions for a central supply monopoly, the Chapter retains a centralising tendency, at least in so far as it implicitly denies to Member States any power to restrict transfers of material within the Community, or imports.[136] Such powers, originally proposed for the Commission, are placed in the hands of the Council, but it seems to be assumed that they can be exercised on a once-and-for-all basis (the draft speaks of the Council making "a Regulation")[137] rather than as a matter of continuing regulation. The free trade principles of the nuclear common market are to remain paramount,[138] and are to be reinforced by the explicit application of the EEC Treaty's competition policy rules.[139] The rules of the Chapter, and of international agreements concluded by the Community, are directly enforceable by the Commission, which may fine persons and undertakings breaching them;[140] supply contracts which are inconsistent with the

[132] *Annex, supra* note 131, arts. 53 & 54.

[133] *Id.* arts. 55-57.

[134] *Id.* arts. 58-63.

[135] *Id.* arts. 64-71; the 1982 proposal retained for the Agency the role of policing contracts for conformity with Community law.

[136] *Annex* art. 53(1).

[137] *Id.* art. 53(2).

[138] *Id.* art. 53.

[139] *Id.* art. 54.

[140] *Id.* art. 74.

rules are declared void.[141] The main roles of the Council, in this scheme, are to determine the substance of the basic control regulation, to supervise Commission activity in such areas as the conclusion of Community supply agreements, and to legislate for situations of supply difficulty.[142] Were the draft, even as revised, to be adopted and operated according to its terms, it would clearly involve more extensive central powers over the working of the market than are contained in the EEC Treaty.

B. The ECSC Regime for Coal

The Framers of the European Coal and Steel Treaty had been a good deal less centralising in their approach. No doubt they did not see security of supply, particularly from outside the Community, as a paramount concern. Development and rationalisation of the Community's own production was far more important, and for this, they trusted largely — though not wholly — to the stimulation that free internal trade would bring. The ECSC Treaty, in fact, does not establish a true customs union with a common external tariff and unified customs procedures. Indeed, it states that

> the powers of the Governments of Member States in matters of commercial policy shall not be affected by this Treaty, save as otherwise provided therein.[143]

It does then provide otherwise to a not inconsiderable extent: minimum and maximum rates of customs duties as against third countries may be established; the High Authority has a limited right of supervision in relation to the administration by Member States of import and export licences and the conclusion of commercial agreements, with powers to ensure that licence arrangements are not unnecessarily restrictive and that such agreements do not hinder the implementation of the Treaty; and the High Authority is also given limited special powers to combat dumping, subsidised imports, and even imports without such vices which arrive "in relatively increased quantities and under such conditions that these imports cause or threaten to cause serious injury to production within the common market of like or directly competing products." Together these provisions impose considerable *potential* restrictions on the freedom of Member States to manage their external trade.[144]

141 *Id.* art. 72.

142 *Id.* arts. 53(2), 61 & 62. *See also* the revised versions of arts. 73 & 74.

143 ECSC Treaty art. 71, para. 1. This freedom, and the conclusions drawn from it by the Court of Justice in Joined Cases 9 & 12/60, Société Commerciale Antoine Vloeberghs v. High Authority, [1961] E.C.R. 197, have led at least one commentator to suggest that the ECSC partakes more of the character of a free trade area than of a customs union: Samkalden, 1961 S.E.W. EUROPA 150, 154.

144 P. REUTER, LA COMMUNAUTÉ EUROPÉENNE DU CHARBON ET DE L'ACIER, ch. 4 (1953).

In fact, if not always in form, this freedom has prevailed over the detailed mechanisms of collective action. The fall in demand that plunged the coal sector into crisis in 1958 provoked individual measures by Member States, later authorised or adopted by the High Authority, rather than a genuine collective response.[145]

This individual approach to a common problem was the more disappointing in that the ECSC Treaty does provide the High Authority with explicit and quite far-reaching powers to tackle situations of glut and scarcity. It is still instructive, now, to look at the operation (or rather non-operation) of these powers in the 1950s, because the story manifests the very considerable difficulty experienced by European states in centralising their response to energy-related resource problems, even in the period when the European idea was thought to be riding higher than it is today and in the presence of explicit Treaty powers in this respect. Let us look first at the arrangements for scarcity, which were the first to become relevant, and then at those for glut.

At the time when the ECSC Treaty was drafted, coal was in short supply in Western Europe. An elaborate set of arrangements for coping with the problem was, therefore, included in the Treaty.[146] With the exception of a three-month period in 1953, they have never been used. Once a serious shortage of coal, too difficult for ordinary Community instruments to deal with, has been found by the High Authority to exist, then, provided the Council does not unanimously decide otherwise, the High Authority may make proposals. At this point one of two things may happen. The Council may unanimously agree on consumption priorities and the allocation of coal resources to the coal and steel industries themselves, to exports and to other sectors of consumption, and the High Authority may then impose production programmes on production enterprises. This is a highly "integrated" solution, in that the measures are taken on a Community level and without explicit regard for national boundaries. The Council, however, may not be able to reach agreement on priorities. In this second case, the High Authority is empowered itself to allocate the resources of the Community among the Member States on the basis of consumption and exports, irrespective of the place of production. It is left to Member States to allocate these resources internally, subject to respect for their export obligations within the Community arising from the allocation scheme and to consulting the High Authority concerning the portions to be allotted to export and to the coal and steel industries themselves. Here the solution is less "integrated": the Member States retain very considerable freedom of action, and the Community measures achieve their effect only

[145] *See generally* D. Blondel-Spinelli, L'Energie dans l'Europe des Six, esp. § 127-37 (1966), and for details of the measures taken in the sphere of external affairs generally, *see* 2 R. Quadri, R. Monaco & A. Trabucchi, Trattato istitutivo della Comunità Europea del carbone e dell'acciaio. Commentario CECA 1004-99 (1970).

[146] ECSC Treaty art. 59; for detailed analysis of this article, *see* Gori, in 2 R. Quadri, R. Monaco & A. Trabucchi, *supra* note 145, at 759-75.

through the action of the States (though it should be noted that the original allocation between States itself represents a decision of great political weight, and a power of a kind which States have been unwilling to grant to the Commission alone within the framework of the EEC Treaty).[147] In both cases, however, the High Authority retains responsibility for allocating resources among undertakings in the coal and steel industries themselves. A further power exercisable by the High Authority with the assent of the Council (for which unanimity is not required) is the restrictions of exports by all Member States to third countries.

As noted above, these powers have been used only once, when the High Authority allocated supplies between States for the first quarter of 1953 to avoid disturbances in the pattern of trade following the opening of the common market.[148] Since 1958 the Community's problem has been one of over-supply rather than under-supply, but it is noteworthy that between 1953 and 1958, despite very considerable supply difficulties stemming from increases in demand, particularly when the Suez Canal was closed in 1956, the High Authority never resorted to these procedures. Instead it preferred, in 1957, to secure the elaboration, on a non-regulatory basis, of consignment plans by the principal producers, with the aim of promoting, *de facto*, an equitable apportionment of Community supplies. It also authorised *de facto* apportionment of West German coal resources by the three sales agencies through which German production was distributed, but its decision to this effect was annulled by the Court of Justice on the ground that the authorised system involved an excessive restriction of competition.[149] This choice of informal means may perhaps have been based on the instruction to the High Authority, in article 57 of the Treaty, to give preference to the indirect means of action at its disposal, such as cooperation with governments to regularise or influence general consumption, or use of its powers on prices and commercial policy.

The difficulty of operating central powers within the Community became clear when shortage turned to glut in 1958 and thereafter.[150] The Community powers to deal with such a situation are contained in article 58, which provides that if the High Authority considers that the Community is confronted with a period of "manifest crisis" which it cannot deal with under article 57, it shall, with the assent of the Council, establish a system of production quotas, accompanied as necessary by import restrictions under article 74(3). If the High Authority does not act, it may be required to do so by a unanimous decision of the Council, acting at the instance of a Member State. The actual determina-

147 Compare the oil allocation arrangements described *infra*, Ch. III, pp. 138-40, which involve both Commission and Council.

148 *See* Gori, *supra* note 146.

149 Case 18/57, Nold v. High Authority, [1959] E.C.R. 41.

150 On the sequence of events in this period, *see* J.F. BESSON, LES GROUPES INDUSTRIELS ET L'EUROPE: L'EXPÉRIENCE DE LA CECA 542 (1962); N.J.D. LUCAS, ENERGY AND THE EUROPEAN COMMUNITIES 28-34 (1977); D. BLONDEL-SPINELLI, *supra* note 145, at 127-40.

tion of quotas is left to the High Authority, on the basis of studies made with undertakings and their associations. It is to be done on an equitable basis, having regard to the basic princples of the Community as set out in articles 2, 3, and 4.

When demand declined in 1957, Member States reacted with individual measures, in particular for control of third-country imports, either through physical restriction, as operated by the French through their coal import monopoly ATIC, or through import duties, as imposed by the Germans in 1958 on all third-country imports in excess of 5 million tonnes per annum. Member States refused, in October 1958, to accept a High Authority proposal for harmonisation of import policies, on the ground that this would infringe their sovereignty. Notice the contrast here with the EEC regime just negotiated by the same six countries (but not, of course, applying to coal) under which a common policy towards imports was to be the norm and individual action would require a derogation.[151] The High Authority, which was finding it difficult even to obtain adequate information from Member States on third-country imports,[152] was relegated to a role of validating their unilateral import measures,[153] including authorisations of frontier inspections by Belgium and Germany to check whether coal imports were of ECSC or third-country origin. In addition, in October 1958 the High Authority secured from Member States acceptance of a temporary scheme for financing stockpiles of coal, partly from Community and partly from Member State funds.

These measures, by early 1959, were clearly proving insufficient, and the High Authority was led to seek from Member States approval for the declaration of a state of manifest crisis, whereunder production and import controls would have come into operation on a Community basis, under articles 58 and 74. Approval was sought on two occasions, in March and May 1959, but failed to obtain the necessary majority in the Council, being opposed by France and Germany on essentially nationalistic grounds. In result, the High Authority secured, at the Community level, only the adoption of a regime for Belgian coal which effectively involved its isolation from the rest of the ECSC and permitted the payment of subsidies on a temporary basis. The High Authority was thus seen to be effectively unable to exercise its ostensibly substantial powers for crisis situations, and was led to tolerate the continuation of national measures on imports, and later, to preside over the institutionalisation of a system of national subsidies for the European coal industry, on which that industry survives today.

It is important to remember that this crisis was being played out at a time when Member States were supposedly showing great political will towards in-

[151] On the actual working out of this regime, *see infra* p. 136.

[152] ECSC High Authority, Seventh General Report on the Activities of the Community, para. 63 (1959).

[153] *See* Recommendation to West Germany of 28 Jan. 1959, authorising the West German Government to impose a tariff quota on imported third-country coal, 1959 J.O. 197.

tegration in the early operation of the EEC. It thus continues to serve as a warning signal of the problems of operating a Community policy in conditions of economic difficulty, even when the general political climate is favourable, let alone in a time when, it is widely agreed, political will has declined to a low level.[154] Clearly, the crisis had an EEC aspect, in that it was cheap imports of oil, as well as of third-country coal, that were cutting the feet from under the Community coal industry; and one reaction to the crisis was the creation of the Community's first series of studies of energy policy as such within an Interexecutive Working Party on Energy.[155] This is not, however, the place to chronicle the long and essentially negative story of the search for a viable Community energy policy, a story well told in several other places.[156] Here we will investigate only how the structure of the EEC Treaty, and the powers it granted, or failed to grant, to the organs of the Community, have shaped the degree and nature of the exercise of Community powers within the framework of energy *supply* policy.

C. The EEC Regime for Oil and Gas Supplies

The belief of the Framers of the EEC Treaty appears to have been that the application of the general principles of the Treaty would be largely sufficient to cope with any difficulties in the supply situation of the energy resources covered by the Treaty: oil, gas, and electricity. Outside the agricultural sector, problems of glut are referred to only obliquely, in provisions on dumping (article 91), on deflection of trade (article 115), and on protective measures for the transitional period (article 226), while problems of shortage are considered briefly in article 103(4), which we return to later. None of these provisions is specific to energy resources.

We need say nothing here about the way in which the EEC Treaty seeks to implement the principle of free trade within its Member States,[157] but it is important to mention the external relations provisions of the Treaty, both because the major sources of EEC energy supply continue, notwithstanding North Sea oil discoveries, to be third-country imports, and because the Treaty, in this respect, is a good deal more rigorous and centralising than Euratom

[154] Stein, *The European Community in 1983. A Less Perfect Union?*, 20 C.M.L. Rev. 641-56 (1983).

[155] N.J.D. Lucas, *supra* note 150, at 34.

[156] *See, e.g., id. passim*; and the highly informed articles by De Bauw, *Oil Policy Before the Yom Kippur War*, in The European Alternatives 79-96 (G. Ionescu ed. 1979), and *La Politique énergétique*, in Les Communautés européennes en fonctionnement/The European Communities in Action 309-32 (D. Lasok & P. Soldatos eds. 1981).

[157] *See supra* Ch. II, *passim*.

or ECSC. The EEC Treaty sets up a customs union, involving the adoption both of a common external tariff[158] and of a common commercial policy

> based on uniform principles, particularly in regard to changes in tariff rates, the conclusion of tariff and trade agreements, the achievement of uniformity in measures of liberalisation, export policy and measures to protect trade such as those to be taken in case of dumping or subsidies.[159]

Under this scheme, Member States abandon the essence of their commercial policy competences to the Community, and have no freedom to take unilateral measures, such as the imposition of temporary national quotas or tariffs on third country imports, save in exceptional circumstances closely defined by the Treaty.[160]

Unfortunately, the internal regime of the EEC does not, as we have seen, appear to have worked very forcefully in relation to energy resources, particularly petroleum,[161] and the external regime of the commercial policy has hardly worked at all. Petroleum products were the last for which levels of duty were agreed in the negotiations on the common customs tariff, in 1964, four years after the level of the great majority of duties had been settled.[162] Crude oil, natural gas and petroleum products are still not explicitly covered by the common regime for third country imports;[163] nor do the common rules for determining the country of origin of products apply to them.[164] The result is that Member States may apply different import policies — for example, one State may impose quotas on petroleum product imports, while others do not — and different tests for determining the origin of such products. Such differences may in turn impair the functioning of the internal common market: a Member State that imposes quotas may, in order to prevent evasion of its policy, be able to invoke a safeguard provision of the Treaty (article 115) so as to prevent imports via other Member States of products that it defines as being of third-country origin.[165] This is contrary to a central principle of the operation of customs unions: free circulation of goods properly imported into the union through any Member State. A similar situation exists in relation to exports: the Community rule of liberalisation of exports still does not cover exports of these products.[166]

158 Arts. 18-29.
159 Art. 113(1).
160 *See* art. 115.
161 *Supra* Ch. II, at pp. 35-44, 56-64 & 77-85.
162 Council Decision (EEC) 64/303 of 8 May 1964, 1964 J.O. 1209 (21 May 1964).
163 *See* Council Regulation (EEC) 926/79 of 8 May 1979, O.J. L 131, 29 May 1979, p.15, art. 17(7) of which imposes very limited constraints on the freedom of Member State action in this field.
164 *See* Council Regulation (EEC) 802/68 of 27 June 1968, J.O. L 148, 28 June 1968, p.1, 1968 O.J. (spec. Eng. ed.) 165, Annex 1.
165 For an example *see* Commission Decision (EEC) 74/271 of 6 Mar. 1974, O.J. L 143, 28 May 1974, p. 15.
166 *See* Council Regulation (EEC) 2603/69 of 20 Dec. 1969, J.O. L 324, 27 Dec. 1969, p. 25, 1969 O.J. (spec. Eng. ed.) 590, *as amended by* Council Regulation (EEC)

The causes of this remarkable failure to bring a group of products within the regime of the Treaty are to be found both in the degree of importance attached by all Member States to oil as a strategic product, and in the strongly divergent approach of Member States to questions of oil policy, best exemplified by the contrast between the dirigiste attitude of the French, manifested particularly through their strenuous adherence to their monopoly import regime, and the free market philosophy pursued in this field by the Dutch and the West Germans;[167] and the further contrast, manifested after the 1973-1974 oil crisis, between the French preference for bilateral arrangements with producer states and the preference for consumer solidarity within the framework of the International Energy Agency shown by all other Member States.[168] In the oil sector, these differences seem to have been a far more significant source of discord, and of inability to operate fully the EEC regime, than have major objective differences such as that between the oil and gas producers and consumers within the EEC. Problems in applying the general Treaty regime have not been removed or even palliated by the development of a "tailor made" energy policy which would recognise the special characteristics of the product. Time and again the Commission has presented to the Council packages of proposals identifying possible Community energy goals and creating instruments for their attainment. Their fate has been well described by a senior Commission energy official:

> Les Etats membres demandent à la Commission de leur proposer un ensemble cohérent de mesures formant l'assise de pareille politique. Mais, dès que la Commission donne suite à cette invitation, le Conseil se montre incapable de prendre une décision de principe opérant les arbitrages nécessaires entre les intérêts divergents des Etats membres; et il indique son souhait de recevoir des propositions formelles, point par point, sur lesquelles il se prononcera. Lorsqu'en revanche la Commission propose une mesure concrète spécifique, l'adoption de celle-ci est fréquemment retardée si pas bloquée, sous le prétexte qu'elle ne peut être décidée sans considération d'un ensemble de mesures plus larges.[169]

1934/82 of 12 July 1982, O.J. L 211, 20 July 1982, p.1. The question whether this exclusion leaves Member States with freedom to follow whatever export policy they wish has been referred to the European Court of Justice — *see* Bulk Oil (Zug) AG v. Sun International Ltd., [1984] 1 W.L.R. 147, [1984] 1 All ER 386 (CA); [1983] 1 Lloyd's Rep. 655 (Bingham J.) — which has decided the question in the affirmative, subject to a requirement of consultation with the Commission: Case 174/84, Bulk Oil (Zug) AG v. Sun International Trading Co. Ltd., (Judgment of 18 Feb. 1986, not yet reported).

[167] *See* Levy, *Relations Between Oil Companies and Consumer State Governments*, 2 J. ENERGY & NAT. RES. L. 9 (1984); T.C. DAINTITH & L. HANCHER, ENERGY STRATEGY IN EUROPE: THE LEGAL FRAMEWORK (1986).

[168] N.J.D. LUCAS, *supra* note 150, at 62.

[169] De Bauw, *La Politique énergétique, supra* note 156, at 313-14: "The Member States ask the Commission to propose a coherent set of measures which might form the basis of a common energy policy. But as soon as the Commission responds to this invitation, the Council is seen to be incapable of taking a decision of principle

These Commission initiatives have not been entirely without result, in the sense that it has been possible for the Council to agree on a number of specific measures in relation both to energy supply and demand. On the supply side, we might mention regulations providing for the reporting of crude oil and natural gas imports[170] and exports[171] and of petroleum product imports[172] (which provide a substitute for an element of the common commercial policy in this area) and for reporting of investment projects in the oil, natural gas and electricity sectors;[173] and directives requiring the reporting of the prices of crude oil and petroleum products.[174] Of greater importance, though, because they show sufficient coherence to be regarded as a part, at least, of a hypothetical common energy policy, are the measures taken, largely under article 103(4) of the EEC Treaty, to deal with shortages of supply.

Obviously the Framers of the EEC Treaty, carrying out their work against the background of the Suez crisis, were not indifferent to the possibility of shortages of materials that would affect the working of the general common market they were constructing. The Six were members of the Oil Committee of the OEEC, which at this time recommended an agreement on emergency oil sharing arrangements within the organisation.[175] At the same time, experience in the ECSC may already have suggested that there were difficulties in legislating in detail for shortage situations, even in a market confined to two principal commodities. We consequently find that the EEC Treaty displays a certain diffidence on this issue. Its nearest approach to envisaging a general Community competence to deal with shortages appears in article 103, on what is called "conjunctural policy" — an inelegant term which is to be understood as referring to the short-term economic policies of the Member States. This article is in vague terms. It enjoins the Member States to "regard their conjunctural policies as matters of common concern", and to "consult each other

involving necessary choices among the divergent interests of the Member States; so it asks for formal proposals on each separate point, on which it may decide. When, however, the Commission proposes a specific and concrete measure, its adoption is frequently delayed, if not blocked altogether, on the pretext that a decision cannot be taken save in the context of consideration of a group of broader measures."

170 Council Regulation (EEC) 1055/72 of 18 May 1972, J.O. L 120, 25 May 1972, p. 3, 1972 O.J. (spec. Eng. ed.) 462, *applied by* Commission Regulation (EEC) 1068/73 of 16 Mar. 1973, O.J. L 113, 28 Apr. 1973, p.1.

171 Council Regulation (EEC) 388/75 of 13 Feb. 1975, O.J. L 45, 19 Feb. 1975, p.1, *applied by* Commission Regulation (EEC) 2678/75 of 6 Oct. 1975, O.J. L 275, 27 Oct. 1975, p.8.

172 Council Regulations (EEC) 3254/74 of 17 Dec. 1974, O.J. L 349, 28 Dec. 1974, p. 1; 649/80 of 17 Mar. 1980, O.J. L 73, 19 Mar. 1980, p. 1; Commission Regulation (EEC) 713/80 of 26 Mar. 1980, O.J. L 81, 27 Mar. 1980, p. 15.

173 Council Regulation (EEC) 1056/72 of 18 May 1972, J.O. L 120, 25 May 1972, p. 7, 1972 O.J. (spec. Eng. ed.) 466.

174 Council Directive 76/491 of 4 May 1976, O.J. L 140, 28 May 1976, p. 4, *applied by* Commission Decision 77/190 of 26 Jan. 1977, O.J. L 61, 5 Mar. 1977, p. 34.

175 N.J.D. LUCAS, *supra* note 150, at 34-35.

and the Commission on the measures to be taken in the light of the prevailing circumstances". It envisages Community action only in that "the Council may, acting unanimously upon a proposal from the Commission, decide upon the measures appropriate to the situation", which measures may, where necessary, be put into effect through Council directives. Its interest for us lies in its concluding provision that its procedures (that is, of consultation, decision and directives) "shall also apply if any difficulty should arise in the supply of certain products." While this article may be employed in circumstances of penury analogous to those referred to in article 59 of the ECSC Treaty, its activation need not involve the creation or exercise of the central powers of control described in that article.

Under article 103(4) there has in fact gradually developed a system for coping with shortages of crude oil and petroleum products which, while not yet severely tested, may prove to be more practicable than the detailed schemes set out in the Euratom and ECSC Treaties. The system borrows heavily from ideas developed by the OECD, but (formally at least) stands entirely on its own and is enforced through EEC machinery. It has four elements, each of which originates in a Council Directive or Decision, and which are here treated in chronological order of adoption.

First, under Directives of 1968 and 1972, Member States are obliged to maintain oil stocks equivalent to ninety days' normal consumption.[176]

Second, under a further Directive of 1973,[177] Member States are obliged to equip themselves with powers to draw on such emergency stocks (which, of course, may be maintained not by the State itself, but by oil companies), to restrict consumption and direct it to priority uses, and to regulate prices to prevent abnormal price increases. It should be noted that this Directive imposes no obligation on Member States to use the powers so obtained in any given circumstances or in a co-ordinated way. Co-ordination was in fact sought through the setting up of a consultative group of their representatives, capable of meeting urgently with the Commission to discuss necessary steps in the event of oil crises, and which the Commission is required to consult at most stages of the procedures outlined below. Practice in the use of the powers, particularly in regard to price control, has in fact varied significantly as between one Member State and another since the passing of the Directive.[178]

Third, since 1977[179] the Commission has been empowered, where difficulties arise in the supply of crude oil or petroleum products, to introduce a system

[176] Council Directive (EEC) 68/144 of 20 Dec. 1968, J.O. L 308, 23 Dec. 1968, p. 14, *as amended by* Council Directive (EEC) 72/425 of 19 Dec. 1972, J.O. L 291, 28 Dec. 1972, p. 154, O.J. (spec. Eng. ed.), 28-30 Dec. 1972, p. 69.

[177] Council Directive (EEC) 73/238 of 24 July 1973, O.J. L 228, 16 Aug. 1973, p. 1.

[178] *See* Levy, *supra* note 167, at 23-28.

[179] Council Decision (EEC) 77/186 of 14 Feb. 1977, O.J. L 61, 5 Mar. 1977, p. 23, *as amended by* Council Decision (EEC) 79/879, O.J. L 270, 27 Oct. 1979, p. 58 *and applied by* Commission Decision (EEC) 78/890 of 28 Sept. 1978, O.J. L 311, 4 Nov. 1978, p. 13.

of licensing of exports of crude oil and petroleum products as between Member States, either on its own initiative, or at the request of a Member State. Initially, the purpose of such licensing is for surveillance of trade between the Member States, but export licences from a particular State may be suspended with the authorisation of the Commission, either (first case) when an actual or imminent shortfall in supply creates an abnormal increase in trade in petroleum products between Member States, or (second case) when a shortfall is likely seriously to endanger a Member State's supplies. A Member State may also (third case) act unilaterally to suspend licences if faced with a sudden crisis where delay would be gravely prejudicial to its economy. Each kind of suspension decision may be referred to the Council, and in the first and third case will lapse after ten days unless approved by it. When the situation, in the opinion of the Commission, no longer justifies the continued application of any of these measures, it may remove them or propose their removal to the Council.

Fourth, again since 1977,[180] where supply difficulties arise as above, the Commission may also set a Community target of a reduction of consumption of up to 10 percent. This decision is applicable for a maximum of two months, after which the Commission may propose for the approval of the Council a further period of reduction with differentiation between the targets for different States according to the extent to which they are capable of substituting other fuels for oil (essentially in electricity generation). These differential savings in oil are to be reallocated from States with large substitution possibilities to those with smaller ones. Alternatively, in the event of a larger shortfall, the Commission may propose a reduction target greater than 10 percent and extend it to other sources of energy.

The degree of Community solidarity evidenced by the measures outlined above may be deceptive. The cynic may doubt whether the Community would have found itself able to take the last of the steps mentioned if the International Energy Agency (IEA), a body which includes all the Community countries save France, together with a number of other consumer nations such as the United States and Japan, had not already brought into operation an international agreement for the reduction of consumption, and the reallocation of supplies, in the event of an oil supply emergency.[181] The system worked out under this agreement is one of considerable elaboration, which operates qua-

[180] Council Decision (EEC) 77/706 of 7 Nov. 1977, O.J. L 292, 16 Nov. 1977, p. 9, *as applied by* Commission Decision (EEC) 79/639 of 15 June 1979, O.J. L 183, 19 July 1979, p. 1.

[181] *I.e.* the International Energy Programme. In 1974, 16 OECD countries (excluding France) signed the Agreement on an International Energy Programme (IEP), to which five other states have since acceded. The IEP deals with every major international energy policy issue which is relevant to western industrialised countries including information gathering and planning for an energy emergency. This latter work is undertaken by the International Energy Agency which was constituted by the IEP as an autonomous body forming part of the OECD and having its headquarters in Paris.

si-automatically when the shortfall in supplies, to the whole group of members or to individual members, exceeds a given level. In the event of a supply crisis, it is the IEA system whose operations would in practice determine the supply position of Member States, not least because the oil companies, whose effective co-operation is essential to any such scheme, are actively associated with the IEA system. The link appears, for example, in the fact that the Commission will only operate the second case of export licence suspension when a Member State's supply shortfall reaches 7 percent,[182] which is the IEA trigger figure. This is not to say that the EEC system is a mere carbon-copy of the IEA one: it has original features, such as the differentiation between substitutable and non-substitutable oil consumption (though this is unlikely to be of great importance in practice), and the device of export surveillance, which can be brought into play in "sub-crisis" situations and has in fact been operating in relation to various interchanges between specific Member States between 1979 and the present time.

Despite its practical dependence, in certain respects, on these other, wider arrangements, it now seems fair to speak of a Community system for oil supply emergencies. While that system represents a substantial achievement, it is worth stressing how it differs, in fundamental respects, from the designs of the Euratom and ECSC Treaties. In this system there is no effective central executive power. The obligations on which its operation depends are wholly imposed on the Member States, and there are no possibilities of direct control by any Community organ over the behaviour of these operating in the oil market.

An important element of the system, moreover, is the re-erection, in the shape of export licences, of precisely those barriers to trade between Member States which it was the general purpose of the Treaty to remove. This can hardly be regarded as a positive step, though it is doubtless a lesser evil than the unilateral and uncoordinated restrictions of trade in hydrocarbons which Member States, in the absence of such a system, would be tempted to impose and which they might seek to defend by resort to article 36.[183] The reason why the Community has been driven to this kind of "least worst" solution is its inability, already remarked, to make any progress with the construction of genuine common policies on energy supply and prices. Member States remain profoundly divided as to the appropriate degree of regulation of the domestic oil market.[184] France, for example, retains control over virtually all aspects of the market, from crude oil imports, through product prices, to service station locations and price-mark-ups.[185] Italy has stringent price controls. At the other ex-

[182] Commission Decision 78/890, *supra* note 179, at art. 3(3).

[183] *See supra* Ch. II, text accompanying notes 66-72.

[184] For a detailed survey of legal regulation of energy markets in the major Community countries, *see* T.C. Daintith & L. Hancher, *supra* note 167, at ch. 5. For details of national price regulations, *see* Levy, *supra* note 167.

[185] Certain elements of the French retail price control system have been abandoned following an adverse ruling by the Euopean Court of Justice in Case 231/83, Cullet v. Centre Leclerc (Judgment of 29 Jan. 1985, not yet reported), [1985] 3 ECJR 33.

treme, Germany, the Netherlands and the United Kingdom currently have no oil price controls in operation. The maintenance of these different systems within the confines of a common market without (official) customs barriers is bound to produce deflections of trade, as was demonstrated at the time of the oil supply crisis in 1973-1974, when West Germany, with relatively less refinery capacity than almost any other country of the EEC, suffered the smallest shortfall in the supply of refined products. Thanks to the Treaty's prohibition of quantitative restrictions on exports, products responded to rising prices in West Germany by flowing there freely from EEC countries where regulation held product prices down.

Since 1981 the Commission appears to have signalled that it no longer thinks it worthwhile to pursue the quest for a centralised common energy policy for the Community. While it is continuing to press for specific measures to be taken for such purposes as harmonisation of energy taxes and transparency of prices, avoidance of oil market instability, and coherence in relations with third countries, it has accepted that differences in the structure and regulation of national energy industries, and in national energy situations, are too profound to permit the adoption of a common overall approach. The term "common energy policy" has been dropped in favour of "Community energy strategy", and the main thrust of the Commission's activities and powers in this field will, it seems, henceforward be directed to monitoring and reinforcing the progress of Member States, collectively and individually, towards the targets for reduction of energy dependence to which they have from time to time committed themselves.[186] The Commission does not, however, intend to mark out a preferred path towards these targets, and it is not even clear that Treaty obligations themselves will necessarily be held paramount, if they appear likely to obstruct the attainment of such targets. Commission tolerance of United Kingdom and Dutch landing requirements may now be seen as the harbinger of this new approach.

186 The key document is EEC Commission, The Development of an Energy Strategy for the Community, Doc. COM(81) 540/final (2 Oct. 1981). *See also* Audland, *European Community Energy Strategy and Its Legislative Implications*, 1 J. ENERGY & NAT. RES. L. 9 (1973); and for a critical discussion, and information on subsequent developments, T.C. DAINTITH & L. HANCHER, *supra* note 167, at ch. 7.

Chapter Four

Conclusions

I. Methods and Degrees of Integration of Energy Markets

In this final chapter we summarise the detailed information we have presented in the two preceding chapters and draw out, by reference to some general ideas about the different approaches to legal integration in Europe and the United States, both the key elements in their respective processes of energy market integration, and the degree to which the American experience may offer lessons for future European policy in this field. Our approach to this task is essentially qualitative, in the sense that we concern ourselves with how the institutions of integration may operate in situations which we perceive as qualitatively different in the Community and the United States by reason of the existence in the latter of fundamental integrating factors which are as yet absent in the Community: a common language, a common currency, an absence of customs posts. We do not attempt any quantification of the integration of energy markets in the two unions, as a measure either of comparative performance to date or of the distances that remain to be covered before a truly common market is attained. Statistics for such indicators as relative energy prices in different parts of the European Community and the United States are available;[1] but their significance in terms of market integration is too uncertain for them to be useful here. The difficulty of constructing consistent time series from the data available, the presence of a variety of explanatory factors not connected with integration, and the problem of disentangling the effects of energy market differences from those of general economic disparities (such as local differences in purchasing power of currency), combine to demand an elabo-

[1] *See, e.g.,* the following publications of the EC Commission: GAS PRICES 1970-1976 (Eurostat, Energy Statistics (Ruby Series) 1977); GAS PRICES 1976-1978 (Eurostat, Energy Statistics (Ruby Series) 1979); GAS PRICES 1978-1984 (Eurostat, Industry & Services, (Blue Covers) 1984); ELECTRICITY PRICES 1973-1978 (Eurostat, Industry & Services (Blue Series) 1980); ELECTRICITY PRICES 1978-1984 (Eurostat, Industry and Services (Blue Covers) 1985). For the USA *see* MONTHLY ENERGY REVIEW (Energy Information Administration, Supt. of Docs., Washington D.C. 1974—).

rate and doubtless controversial effort of statistical interpretation, which we are happy to leave to others.

Content, therefore, to assume that the European Community energy markets are less integrated after thirty years than those of the United States after two hundred, we here consider Europe's prospects for further progress in the endeavour in the light of comparison of the general integrative apparatus of the Community and of the United States. The most salient features of these apparatus can be quickly summarised. First, the Community Treaties are far more detailed and explicit than is the United States Constitution in their commitment to the creation of a common market.[2] Prohibitions on state restraint of trade which have had to be developed by United States courts from the most general declarations of constitutional principle are to be found specifically expressed in the Treaties: discriminatory state taxation provides an example.[3] Similar solutions have not always been reached: consider the case, already discussed in detail, of state aids and state enterprise, where the European rules are both more specific and more severe.[4] Second, in most sectors the Treaties do not expressly contemplate the exercise of centralised policy-making powers, save in exceptional circumstances such as glut or scarcity[5] or the inadequacy of other Treaty instruments.[6] No such limitations are attached to the use of federal power in the United States, whose constitutional limits are set only by current perceptions of the breadth of the commerce and other general constitutional powers. The legitimating basis of centralised policy in the Community is thus much narrower and more fragile than in the United States. Third, the Council, the Community's central legislative-executive

[2] For a comparison of the US and EC approaches, *see generally* Kommers & Waelbroeck, *Legal Integration and the Free Movement of Goods: The American and European Experience,* in I/3 INTEGRATION THROUGH LAW 165 (1986); Heller & Pelkmans, *The Federal Economy: Law and Economic Integration and the Positive State — The U.S.A. and Europe Compared in an Economic Perspective,* in I/1 INTEGRATION THROUGH LAW 245 (1986); Krislov, Ehlermann & Weiler, *The Political Organs and the Decision-Making Process in the United States and the European Community,* in I/2 INTEGRATION THROUGH LAW 3, 21-22 (1986).

[3] *See supra* Ch. II, pp. 64-85.

[4] *See supra* Ch. II, pp. 44-64. The significance of this specificity and severity, of course, depends on the way in which the rules are enforced. We return to this point, with particular reference to state enterprise in the Community, *infra* pp. 149-52.

[5] *See supra* Ch. III, pp. 132-35 & 138-42.

[6] *See* EEC Treaty art. 235, discussed in more detail in Gaja, Hay & Rotunda, *Instruments for Legal Integration in the European Community — A Review,* I/2 INTEGRATION THROUGH LAW 113, 117-20 (1986); Jacobs & Karst, *The "Federal" Legal Order: The U.S.A. and Europe Compared — A Juridical Perspective,* in I/1 INTEGRATION THROUGH LAW 169 (1986); Krislov, Ehlermann & Weiler, *supra* note 2, at 22; Stein, *Towards a European Foreign Policy? The European Foreign Affairs System from the Perspective of the United States Constitution,* in I/3 INTEGRATION THROUGH LAW 3, 21-22 (1986). For reasons that will appear below, we do not expect to see energetic use of art. 235 in the energy sector in the near future.

body, is a committee drawn from Member State governments. This leads, even where the Treaties do not expressly demand it, to an exercise of central power which is substantively conditioned by Member State situations, interests and practices, and which tends to rely heavily, for its implementation, on national administrations. With certain limited exceptions[7] (of which energy is not one) the key central function in the Community, vis-à-vis the Member States, is thus one of co-ordination, rather than substitution. In the United States, by contrast, the independent constitution of the federal legislative and executive organs leads to direct centralised administration, substituting in its own spheres for state administration, co-ordination with which requires special and unusual efforts. If we regard two essential elements of a legal scheme of market integration as being *control* of disruptive local activity, and *capacity* for central governance of the market, we may say that the Community appears strong in the former, weak in the latter; in the United States, however, the strength is in capacity, while control may be problematical because of the weakness of its constitutional bases.

In the two foregoing chapters, we have tried to adduce the data specific to the energy sector which would illustrate the points just made in summary form. In the next section we use this information to show how far these ideas hold for the energy sector, and in the final section attempt to make some suggestions about the future institutional development of Community policy for energy-related natural resources in the light of the American experience. Before doing this, we would make two general points to avoid misunderstanding. First, it should not be assumed, in connection with the control/capacity contrast, that strength is necessarily a virtue: indeed, the contrary will appear when we come to consider United States federal energy policy. Second, it should not be assumed that the control/capacity coupling is symmetrical: part of our argument will be that while United States control problems are largely rendered irrelevant by the strength of central capacity (allied with the preemption doctrine), the inverse relationship does not hold for the Community, where strength in control (itself more apparent than real) cannot supply the place of weak central capacity.

II. Key Factors for Legal Integration in Energy Markets

A. The State as Resource-Owner and Entrepreneur

In the foregoing remarks about formal differences of legal and constitutional structure, we have had to omit one factual matter which appears to us to be of key importance in relating American and European experiences of integra-

[7] *E.g.* agriculture or competition policy.

tion in the energy sector. This is the issue of the structure of energy markets, and in particular of the role of the state therein: that is to say, the role of the state as market participant. The basic paradigm of legal integration of national markets, as expressed both in the practice of the Supreme Court under the United States Constitution, and in the key articles of the EEC Treaty (9, 16, 30, 34, 95) is one of control of state regulation and taxation of private enterprise. As we have seen, however,[8] many countries, the Member States of the European Communities among them, have not found this to be an adequate or appropriate means for securing the control they desire over natural resources within their territory or the financial returns they seek therefrom. In addition, they have obtained proprietary rights, both over energy-related natural resources *in situ* and, in varying measure, in the processes of their production, processing and sale. In the United States, on the other hand, the most significant instances of public ownership of energy resources *in situ* are federal (notably federal ownership of about 60 percent of all Western coal, all outer continental shelf petroleum, and some onshore petroleum) and thus inherently an integrative factor. The state of Alaska's petroleum ownership is the most significant exception to this pattern. Participation in production through the creation of state oil corporations, or of analogous enterprises for other natural resources, is virtually unknown.[9] This contrast makes itself felt in a number of ways. It helps to explain the absence, in Europe, of a counterpart of the body of regulation of natural resource exploitation designed in the United States to resolve the problems of inadequate specification of (private) rights to the natural resources of oil, gas and water.[10] Where the state owns or wholly controls the resource in *situ* it can, even where it wishes to entrust to others the risks (and profits) of exploitation, do so in a way which avoids property right uncertainties.

The European predilection for a measure of state ownership also explains, of course, the care taken, particularly in the EEC Treaty, to provide rules to govern the relationship of the Member States with their public enterprises and the degree of subjection of such enterprises to the general rules of the Treaty on competition and other matters.[11] These rules, as we have seen, are not lacking in ambiguity, and the Commission's first attempt to concretise them by means of a directive has been controversial:[12] but they do at least clearly enunciate the principle that the same basic Treaty ideas — avoidance of distortion of competition, of discrimination on grounds of nationality, of barriers to intra-Community trade — should apply to the exercise of the state's proprietary

[8] *See supra* Ch. I, pp. 8-15.

[9] *See supra* Ch. I, p. 15.

[10] *See supra* Ch. II, pp. 85-98.

[11] *See supra* Ch. II, pp. 56-64.

[12] Commission Directive 80/723 of 25 June 1980 on the transparency of financial relations between Member States and public undertakings, O.J. L 195, 29 July 1980, p. 35; Joined Cases 188-190/80, French Republic, Italian Republic and United Kingdom v. Commision, [1982] E.C.R. 2545.

powers as apply to the exercise of its powers of taxation and regulation. In the United States, perhaps because of the unfamiliarity of the phenomenon of a state's affecting commerce not by regulation but by trading activity or the granting of financial assistance, the Supreme Court seems to be groping for principles in its decisions in this field.[13] While the decisions are not easy to understand or reconcile, it does at least appear that the Supreme Court is willing to distinguish between the activities of the states as market participants on the one hand, and market regulators on the other hand, and to permit the states to exercise a significantly greater degree of express discrimination in the former activities than in the latter. This tolerance of potentially market-fragmenting behaviour is, we would suggest, only explicable by reference to the relative rarity of the state public enterprise phenomenon in the United States. It is hard to believe that the Supreme Court could maintain a view that the application of the commerce clause to the public sector was too "difficult",[14] in the face of an industrial and commercial public sector of the size and variety of, say, that of Italy or the United Kingdom.

B. The Exercise of the Control Function

The effective implementation of the Treaty articles controlling the behaviour of the state as market participant is of special importance in the energy sector, given the prevalence of state enterprise and the frequent presence, also, of enterprises to which the state grants special or exclusive rights in terms of article 90: holders of petroleum production licences, privately- or locally-owned gas and electricity utilities, licence holders within the framework of the French petroleum monopoly and so on.[15] In fact the level of implementation has been low, and this has seriously weakened the Community's control function in this area. Two modes of implementation need to be distinguished: Commission supervision and enforcement of Member State obligations, ultimately through legal process in the Court of Justice; and private enforcement through national legal process.

In its control structure the Community differs significantly from the United States. The Commission's supervision function has no equivalent in the Constitution. The Constitution provides only that cases arising under it shall be within the federal judicial power, and leaves Congress a large discretion as to the organisation and extent of the exercise of that power:[16] by 1816 a combina-

[13] *See* Hughes v. Alexandria Scrap Corp., 426 U.S. 794 (1976); Reeves, Inc. v. Stake, 447 U.S. 429 (1980); Hicklin v. Orbeck, 437 U.S. 518 (1978). *See supra* Ch. II, pp. 45-56.

[14] 447 U.S. 429, 439 (1980).

[15] For more detail *see* T.C. Daintith & L. Hancher, Energy Strategy in Europe: The Legal Framework (1986).

[16] U.S.Const. art. III, § 2.

tion of Congressional legislation and Supreme Court decision[17] had established a situation whereunder infractions by states of their constitutional obligations might be litigated as far as the Supreme Court by aggrieved persons subject to ordinary rules of standing to sue. In consequence, private initiative became established at an early date as an important element of constitutional protection. Its vigorous development, furnishing the material for the Supreme Court's negative commerce clause and preemption jurisprudence, may partially explain the fact that Congress has not in recent times felt the need to endow federal administrative agencies with powers to police compliance by states with relevant constitutional limitations nor (save on rare occasions) itself to prohibit particular types of parochial legislation by states. In Europe, by contrast, the possibility of enforcement through private initiative was probably not envisaged by the Framers of the EEC Treaty: it has, however, developed as a significant element of the European legal order since the European Court held in 1962 that certain obligations of Member States under Community law could be directly enforced through litigation by individuals in their national courts.[18]

In the light of this development, it is striking that in the thirty-year life of the Communities there should have been only a handful of cases before the European Court directly concerned with Member State measures to restrain trade in energy-related natural resources,[19] none of which raised the "local sovereignty" issues frequently litigated in the United States. Obviously the difference in European and American resource endowments is the major factor here; but this does not explain why the attempts of Oklahoma and West

[17] *See* Judiciary Act of 1789, § 25, and Martin v. Hunter's Lessee, 14 U.S. (1 Wheat.) 304 (1816).

[18] Case 26/62, Van Gend en Loos v. Netherlandse Administratie der Belastingen, [1963] E.C.R. 1.

[19] *I.e.*, Joined Cases 9 & 12/60, Société Commerciale Antoine Vloeberghs S.A. v. High Authority, [1961] E.C.R. 197 (third-country coal imports); Case 6/64, Costa v. Enel, [1964] E.C.R. 585 (Italian electricity nationalisation); Case 20/64, S.A.R.L. Albatros v. Société des Pétroles et des Combustibles Liquides (SOPECO), [1965] E.C.R. 29 (adjustment of French oil import monopoly); Case 7/71, Commission v. France, [1971] E.C.R. 1003 (functioning of Euratom Supply Agency); Ruling 1/78, delivered pursuant to EAEC Treaty Art. 103(3) (Draft Convention of the International Atomic Energy Agency on the Physical Protection of Nuclear Materials, Facilities and Transports), [1978] E.C.R. 2151; Case 72/83, Campus Oil v. Minister for Industry and Energy, Ireland, [1984] E.C.R. 2727 (free movement of goods — supply of petroleum products); Case 231/83, Cullet v. Centre Leclerc Toulouse, (Judgment of 29 Jan. 1985, not yet reported), [1985] 3 ECJR 33 (French minimum petrol price rules); Case 174/84, Bulk Oil (Zug) v. Sun International Trading Co. Ltd. (Judgment of 18 Feb. 1986, not yet reported) (legality of UK oil export policy). At the time of writing two important cases are before the Court under art. 177 procedure: Joined Cases 153 & 156/83, *STA 31 v. Levallois Distribution* and *Rosello v. Sodivar*, both concerning French rules setting minimum petrol prices (O.J. C 222, 19 Aug. 1983, p. 4).

Virginia to reserve state natural gas to their citizens were speedily litigated to destruction[20] whereas the perhaps subtler, but similarly motivated measures by the United Kingdom and the Netherlands[21] have not been attacked in the courts. Certainly the European Commission, as we shall see, has had its own reasons for restraint in this area, but we still need to explain why no would-be exporter of Dutch or British gas, or no disappointed European purchaser, ever went to the courts with a complaint that his legitimate commercial objectives were being frustrated by a national policy in breach of Community law.

One part of the answer to this query lies, perhaps, in the judicial structure of the Communities. Such actions will, as already noted, need to start in a national court, probably the courts of one of the allegedly offending parties, and their reaching the European Court for authoritative interpretation of the Treaty will depend, under the article 177 procedure, on the making of a reference by the national court. Even where such a court has an obligation — as opposed to a discretion — to make such a reference, there exists no obvious machinery for enforcing this obligation. Experience suggests that a litigant might have some reason to feel deterred by fear of a not wholly objective treatment by the national court, either in disposing of the case without a reference to the European Court in order to avoid a ruling from that Court adverse to an important element of national policy,[22] or (less probably) in the implementation of such a ruling where a reference had been made. This procedural barrier, and any deterrence to litigants it creates, have, however, often been overcome.

Additional, and different, barriers result from state participation. If the Member State *owns* the resource *in situ*, the would-be exploiter, even though a private enterprise, will be dependent on the Member State for all future access to the resource under leases or other forms of contract yet to be granted, and may be exposed to state discretionary measures even under existing contracts. These dependencies create a powerful deterrent against any litigation, whether legally promising or not, which may offend the authorities of the grantor state. This deterrent factor is considerably weaker where the state enjoys only regulatory, not proprietary rights. If the Member State itself *exploits* the resource, through some form of public enterprise, the chances of discontent among would-be private exporters diminish as the share of production accounted for by the state enterprise(s) increases. In addition to there simply being less of the resource in private hands, other inhibiting factors may come

[20] Oklahoma v. Kansas Natural Gas Co., 221 U.S. 229 (1911); Pennsylvania v. West Virginia, 262 U.S. 553 (1923). *See supra* Ch. II, text accompanying notes 19-30.

[21] *See supra* Ch. II, pp. 35-44.

[22] *See* the decision of the French Conseil d'Etat, C.E. 19 June 1964, Société des Pétroles Shell-Berre, [1964] Rec. Lebon 344, [1964] Rev. Dr. Pub. 1018 (conclusions Questiaux), [1964] C.M.L.R. 462. But there have been recent references, involving French petroleum policy, from the judicial courts: *see* Cases 153, 156 & 231/83, *cited supra* note 19.

into play (for example, the fact that in the UK sector of the North Sea, all private producers of oil and gas were until recently linked with BNOC or BGC in a formal joint venture, under the state participation arrangements already described).[23] A further inhibition may be created if the principal would-be importers of the natural resource in question are themselves state enterprises, as is often the case with gas-distribution and electricity-generation enterprises in the Community. The association of such enterprises with their own national government, whether real or only formal, introduces a political element into what would otherwise be commercial litigation. The "consumer" state government may well prefer to seek a settlement through diplomatic channels and may prohibit, or at least discourage, such litigation.

A further point is that there may be much greater difficulties in discerning and proving a discriminatory purpose in state trading, as opposed to regulatory, behaviour. While traders need rules to organise their business behaviour and may in many cases announce these rules publicly (price lists, trade terms, and so on), much decision-taking will remain discretionary. Only rarely, as in relation to coal pricing,[24] does European law require the adoption of rulebased selling, by such means as adherence to published price lists. The area of discretionary commercial decision is likely to be largest in times of trading difficulties — shortages of supply, price instability, etc. — and in situations where obligations of nondiscrimination are imposed on enterprises (public utilities, or holders of dominant market positions). What is striking about the *Reeves* case[25] is that the state enterprise adopted its discriminatory selling policy by rule, rather than relying on commercial discretion. Had it simply started turning down out-of-state orders, referring vaguely by way of justification to "shortage" and "priority customers", one wonders if the case would have got as far as it did. Disadvantaged European enterprises may well have been discouraged from litigation by the burden of proving that a specific discriminatory purpose lay concealed behind such vague business formulae.

This essay is perhaps not the place to enter into a comprehensive discussion of the capacity of courts, in a federal system, to act as a motor of integration by limiting states strictly to their assigned or retained competences, thereby creating regulatory "gaps" which must be filled by federal, or at least joint, action if they are to be filled at all.[26] Within the Communities this has, since 1962, been an important process, particularly in such fields as agriculture and

23 *See supra* Ch. II, text accompanying notes 137-54, and *supra* note 137 for the recent changes.

24 ECSC Treaty art. 60.

25 Cited *supra* note 13.

26 On the Regulatory Gap Theory *see generally* Cappelletti, Seccombe & Weiler, *Integration Through Law: Europe and the American Federal Experience — A General Introduction*, in I/1 INTEGRATION THROUGH LAW 3, 21 (1986); Heller & Pelkmans, *supra* note 2, esp. at §§ I & II. For the applicability of the theory in other areas, *see, e.g.*, T. BOURGOIGNIE & D. TRUBEK, CONSUMER LAW, COMMON MARKETS AND FEDERALISM IN EUROPE AND THE UNITED STATES 3-4 (II Integration Through Law Series, 1986).

freedom of establishment (though there have appeared in the last few years clear signs of a limit to its integrative potential).[27] The European Court has, however, hitherto[28] been offered virtually no opportunity to enforce compliance by the Member States with their obligations of free trade in energy-related natural resources, for reasons which, we would suggest, are largely connected with the prevalence of state ownership and participation in this sector of the economy. In consequence of this unusual lack of private initiative, the enforcement burden in this sector has rested almost wholly on the Commission. At first sight it may not look as though the Commission has tried very hard: it has brought only two energy cases before the Court, both under the Euratom Treaty,[29] and has prosecuted (unsuccessfully) only one alleged competition law violation.[30] It took until 1979 to secure the reform of the French oil import monopoly so as to eliminate its main discriminatory and trade-restricting features;[31] the British and Dutch restraints on oil and gas exports, as we have seen, continue in force. In fact, the Commission has not been as inactive, or ineffective, as these bare facts might suggest. Several major investigations have been undertaken into Community petroleum markets;[32] and even if the prosecution of the British Petroleum Company and other major oil companies for alleged abuse of dominant position during the 1973-1974 oil crisis was ultimately unsuccessful, it did, by underlining the need for coherence in Member State reactions to oil shortages, find some later reflection in the elaboration of Community policy for dealing with such shortages. Significant results have also been obtained in the sphere of state aids (articles 92 and 93), with the elimination of Belgian and Dutch aids to refinery construction and Dutch aids to electricity costs in glasshouses (and the acceptance, after examination, of German aids to oil exploration).[33] Other elements of the story, however, can only be explained in terms of a strategic acceptance by the Commission of a second-best situation, in which, in the absence of Community policy

[27] A landmark event here was French defiance of the Court of Justice's ruling against it in the "lamb war" case, Case 232/78, Commission v. France, [1979] E.C.R. 2729 (for the sequel *see* Joined Cases 24 & 97/80R, [1980] E.C.R. 1319). This case may be seen as an example of judicial identification of a regulatory "gap" caused by the inability of the Council (through lack of agreement) to exercise its powers. It may be argued that since the purpose and effect of the French defiance was to get the Council to exercise these powers, the process of integration was forwarded by the decision, but the price, in terms of lost respect for the obligatory character of the judicial decision, may still be too high.

[28] *See* now the references cited *supra* at the end of note 19.

[29] Case 7/71, Commission v. France, [1971] E.C.R. 1003, and Ruling 1/78, *Draft Convention of the International Atomic Energy Agency*, [1978] E.C.R. 2151.

[30] Case 77/77, B.P. v. Commission, [1978] E.C.R. 1513.

[31] EC Commission, Ninth Report on Competition Policy para. 205 (1980).

[32] EC Commission, Sixth Report on Competition Policy para. 268 (1977).

[33] *See* Commission Decision 73/293 of 11 Sept. 1973, O.J. L 270, 27 Sept. 1973, p. 22; Commission Decision 80/1157/EEC, *Re* Investment Aids at Antwerp, O.J. L 343, 18 Dec. 1980, p. 38, [1982] 3 C.M.L.R. 138; Bull. EC 12-1969, p. 28.

instruments such as a common export policy for hydrocarbons, Member States are allowed to maintain national controls which may be used so as to serve Community ends (or at worst, so as to serve national ends without damage to the interests of other Member States), even though such controls are *ex facie* incompatible with the principles of the Treaties. A variety of Commission "tolerances" may be explained in this fashion, from coal subsidies in 1958[34] to the operations of BNOC up to 1985.[35]

C. The Community's Weakness in Central Governance Capacity

The general conclusion one may draw from a survey of energy policy is that the Community's capacity for central governance of energy markets has been so weak that it has been impossible to exercise in the full measure demanded by the Treaties the function of control of disruptive local activity. Hitherto, the Community has been unable *either* to develop a uniform framework for energy supply relationships between its Member States and the rest of the world, a serious failing given the Community's continuing overall situation of energy dependence; *or* to harmonise the fundamental conditions of operation of national energy markets, in terms of such criteria as types and levels of energy taxes, national price policies and controls, restrictions on market entry and so on. The major variations subsisting between national markets — for example, in price levels in times of shortage of oil products, in powers of control of exports, in structures for determining prices charged by energy utilities — mean that due exercise of control powers may create serious risks for Community interests or may be impossible for lack of information. A country like Italy, for which price control of some basic commodities like petroleum products is an important part of a delicate political balance, is less likely to abandon that control than to defy an attempt to enforce free movement of petroleum products to the richer and freer German market: hence the adoption in 1977 of Commission-supervised market segmentation in times of oil supply difficulty. Until the Community has a common export policy for hydrocarbons, removal of the United Kingdom and Dutch destination controls might leave the Community defenceless against leakage of its own supplies in times of oil crisis. Identification of competition-distorting aids to major consumers of gas and electricity is all but impossible so long as the Member States' financial relationships with publicly owned gas and electricity utilities remain in the obscurity afforded by legal regimes for public ownership which differ in important respects and are hard to compare.[36] This last example reminds us once

[34] *See supra* Ch. III, pp. 133-35.

[35] *See supra* Ch. II, pp. 59-63.

[36] A first attempt to get round this problem is represented by the Commission's "transparency" directive, Directive 80/723, *supra* note 12. This directive does not apply to energy undertakings.

again of the all-pervading importance of public enterprise in explaining the lack of progress of integration in the energy sector.

In disclosing a situation where a weakness of capacity engenders, in its turn, a weakness of control, the Community energy sector presents a contrast both with most other sectors of the Community's economic life and with the United States situation. In other sectors, the relationship between control and capacity has generally been thought to be one of compensation, in the sense that weaknesses in central governance capacity have to some extent at least been palliated by the forcefulness of control activity, often privately initiated and pursued through the European Court with no more than encouragement from the Commission. As national regulatory regimes have been rendered legally insecure by European Court judgments, so the search, it is argued, for Community solutions where regulation is deemed essential is correspondingly reinforced. As we have already noted, the problem of the energy sector may be that by reason of its high degree of state and public participation, it lacks the Toepfers, Lüttickes and de Peijpers[37] who step in where the angels of the Commission fear to tread.

D. United States Strength in Central Governance Capacity

We should also note the contrast with the dynamics of the United States control/capacity relationship. There the strength is in the central governance capacity of the federal executive and legislative organs, which enjoy an independently legitimated source of power, not passing through the governing institutions of the states. Federal power is not only independent: it is also very extensive. Following Supreme Court interpretations of the commerce clause in the 1940s,[38] it is hard to envisage aspects of the energy industries that could not be the subject of federal regulation under the commerce clause. In addition, the Federal Government holds massive proprietary powers, in particular over onshore coal and offshore oil and gas reserves.

The essentially unlimited character of federal legislative power tends to reduce the importance of any weaknesses in courts' ability to control parochial state behaviour. Congress can, for example, readily control state efforts to export energy tax burdens by setting ceiling levels for such taxes. Similarly, it

[37] Toepfer: *see* Joined Cases 106-107/63, [1965] E.C.R. 405; Joined Cases 5, 7 & 13-24/66, [1967] E.C.R. 245; Case 250/80, [1981] E.C.R. 2465.

Lütticke: *see* Joined Cases 31-33/62, [1962] E.C.R. 501; Case 48/65, [1966] E.C.R. 19; Case 57/65, [1966] E.C.R. 205; Case 4/69, [1971] E.C.R. 325; Case 51/70, [1971] E.C.R. 121; Case 42/72, [1973] E.C.R. 57.

De Peijper (or Centrafarm): *see* Case 24/67, [1968] E.C.R. 55; Case 15/74, [1974] E.C.R. 1147; Case 16/74, [1974] E.C.R. 1183; Case 104/75, [1976] E.C.R. 613; Case 107/76, [1977] E.C.R.957; Case 102/77, [1978] E.C.R. 1139; Case 3/78, [1978] E.C.R. 1823.

[38] Wickard v. Filburn, 317 U.S. 111 (1942).

could respond effectively and without serious constitutional constraint if, in important fields like energy, states were to turn to public enterprise as a means of avoiding judicial restrictions on their regulatory power. Finally, Congress has the option of adopting its own affirmative scheme of substantive regulation, with the justifiable assurance that, as a side effect, courts would invoke the preemption doctrine to invalidate any conflicting state activity in the same field.

On this view, the Supreme Court/Congress relationship may be quite different from that suggested by the theory of regulatory "gaps", of which, as already noted, we have found no clear evidence in our examination of the development of federal law regulating the natural gas industry.[39] The theory of the regulatory gap is that Congress is compelled to act by supplying national regulation after the Supreme Court has invalidated the basis on which state regulations rested. It seems at least equally plausible to argue that the availability of Congressional regulatory power may lead the Court to be cautious in striking down state schemes of regulation, taxation or market participation, in the knowledge that, given sufficient political demand, Congress can substitute its own solution. Obviously there is no necessary inconsistency between these two ideas. Both types of influence may have operated in the United States.

We have already indicated our doubts about Congress' judgment in the exercise of its ample regulatory powers. It succeeded in shifting wealth away from energy producers, and, of the total that it denied producers, it transferred a modest fraction to consumers. But these transfers were achieved at high cost, including: (1) loss of available domestic energy that could have been produced at costs at or below what consumers were willing to pay; (2) use of resources to secure substitutes (*e.g.*, imported oil and natural gas) at prices exceeding both the value to consumers and the cost of supplies whose production the price controls had foreclosed; (3) unemployment of labour and capital in enterprises that were interrupted by natural gas curtailments; (4) waste of valuable time in the queueing that was necessary to clear retail gasoline markets when federal bureaucracies failed to match supply with demand; (5) in a final irony, the balkanisation of energy markets through federal freezing of regional energy distribution patterns. The costs of these programmes seem particulary high when one recognises that the redistributive achievements — however one may value them — could have been achieved far more cheaply through rent-capturing taxes.

Thus the United States adopted solutions that were "integrated" in the technical sense of being imposed by central government without impediment by the states. It is hard, however, to view the United States performance as living up to a more substantive concept of integration — production and distribution of energy in such a way as to maximise aggregate welfare. If the states had been left juridically free to pursue their own policies and constrained only by the market's natural limitations on rent-seeking activity by small jurisdictions, it seems doubtful whether they could have done worse.

[39] *See supra* Ch. III, pp. 101-13.

III. The Future Style of Community Action in the Energy Sector

In the energy sector, mistakes like those of the United States appear to be beyond the present reach of the Community. The bases for centralised action provided by the Treaties are narrow and specific: emergency measures, tax harmonisation, common commercial policy, etc. Community action even within these limited spheres is still far from complete, and its only significant extension, the erection of the Community's oil emergency policy on the basis of EEC Treaty article 103(4), is designed around the acceptance of non-integrated energy markets. In these circumstances, notwithstanding the formal facilities offered by EEC Treaty article 235, there is no possibility of the Community's adopting its own regulatory or tax programmes in the energy sector in substitution for policies hitherto operated by Member States. Given the performance of the federal authorities in the United States, and the Community's own experience in agricultural policy, this is no tragedy. If, however, one discards complete policy integration as the ultimate goal — and it is clear that this is what the Commission has done — what working principles for securing further integration (if any) can be established?

While maintaining support for schemes of fundamental institutional reform, under which this question would be presented in a quite new light,[40] the Commission has already given an answer in terms of the existing Treaty framework in its 1981 paper entitled "The Development of an Energy Strategy for the Community".[41] This is a startling document, in that it downgrades economic integration — whether understood as control of Member State disruption, harmonisation or centralisation of policy-making — from the rank of an objective whose achievement will be productive of economic and social welfare[42] to that of one instrument among others for the achievement of Community energy policy objectives, such as reduction of dependence on energy imports. The paramount Community function in the field of energy policy is seen as the setting of common objectives to which Member States jointly, as well as individually, commit themselves. The progress of Member States towards these objectives will be monitored by the Commission and where necessary reinforced by its invocation of Treaty controls; but Member States may

[40] *E.g.*, the Spinelli initiatives for a Treaty of European Union: *see* Draft Treaty Establishing the European Union (and Resolution of the European Parliament thereon), O.J. C 77, 19 Mar. 1984, p. 33, BULL. EC 2-1984, p. 7 (on which *see generally* AN EVER CLOSER UNION: A CRITICAL ANALYSIS OF THE DRAFT TREATY ESTABLISHING THE EUROPEAN UNION (R. Bieber, J.P. Jacqué & J. Weiler eds., European Perspectives Series, EC Commission, 1985)).

[41] Doc. COM(81) 540/final (2 Oct. 1981). For a full discussion *see* T.C. DAINTITH & L. HANCHER, *supra* note 15.

[42] *See* EEC Treaty, Preamble, para.2, and art. 2.

choose their own paths, taking account of their diverse energy endowments, market structures and policy preferences. The exercise of Community competences is to be limited to cases where the Treaties so provide or the Community as a whole is better equipped than Member States individually to secure given results: for example, where economies of scale are available in research and development, where joint negotiation with third countries may increase the Community's influence, or where centralised action is otherwise more efficient, as in securing the more transparent operation of the oil market. If the Commission's hope was that this new presentation would prod the Council into swift action in the exercise of a restricted range of Community competences, it has so far been disappointed. The degree of immobility in Community energy policy remains about the same as before 1981.

More interesting is the view of the integration process disclosed by the document. In this view, the potential range of central action is severely limited. While harmonisation of energy taxes remains on the agenda, the watchword in relation to price regimes becomes one of transparency rather than alignment.[43] The persistence of separate national energy markets is thus tolerated, and the "second-best" solution for the exercise of the Community's control function[44] is institutionalised. We should thus not expect the Commission to attack national practices which, *in the present state of integration*, are neutral or positive with regard to energy policy goals, unless important interests outside the energy sector are seriously damaged thereby. In this attitude there is a distorted echo of the American situation. There, as we have argued, local variations and even disruptions can be tolerated because of the existence of comprehensive central power: where unity can be chosen, diversity can also be chosen. In the Community, local diversity and disruption are to be tolerated because the Commission assumes that central power *cannot* be exercised for fear of precipitating a Community crisis, in which some Member States would press for the closing of a perceived "regulatory gap" by Community legislation which others, satisfied with a free trade situation, would block. The eventual result might be a damaging defiance by States in the first group. Where unity is seen as unavailable, diversity becomes a matter of compulsion, not choice. The fact that we would have wished the United States to opt more frequently for diversity in the energy sector implies no satisfaction with the current state of Community energy policy. The Community is still in the process of trying to eliminate or diminish major distortions to trade founded on state regulation and state enterprise within its energy markets, a process of elimina-

[43] *See* Draft Council Resolution, para. 2, in Energy Pricing-Policy and Transparency, Doc. COM(81) 539/final, p. 17 (1 Oct. 1981); Council Recommendation 81/924 of 27 Oct. 1981 on electricity tariff structures in the Community, O.J. L 337, 24 Nov. 1981, p. 12; Council Recommendation 83/230 of 21 April 1983 on the methods of forming natural gas prices and tariffs in the Community, O.J. L 123, 11 May 1983, p. 40.

[44] *See supra* pp. 151-52.

tion long completed in the United States. Only when such distortions are largely removed can we enter a phase of evaluating centralised as against localised, uniform as against diverse, solutions to Community energy policy problems, a phase in which we shall have much to learn from recent United States experience.

References

Books

B. Ackerman & W. Hassler, Clean Coal/Dirty Air (1981).

American Gas Association, Competition in the Natural Gas Industry (14 Feb. 1984).

J.F. Besson, Les Groupes industriels et l'Europe: L'expérience de la CECA (1962).

R. Bieber, J.P. Jacqué & J. Weiler eds., An Ever Closer Union: A Critical Analysis of the Draft Treaty Establishing the European Union (European Perspectives Series, EC Commission, 1985).

D. Blondel-Spinelli, L'Energie dans l'Europe des Six (1966).

D.R. Bohi, Analyzing Demand Behavior: A Study of Energy Elasticities (1981).

T. Bourgoignie & D. Trubek, Consumer Law, Common Markets and Federalism in Europe and the United States (II Integration Through Law Series 1986).

S. Breyer, Regulation and Its Reform (1982).

M. Cappelletti, M. Seccombe & J. Weiler gen. eds., Integration Through Law: Europe and the American Federal Experience (A series in 6 volumes, 1985-1986).

R. Churchill, K.R. Simmonds, & J. Welch eds., New Directions in the Law of the Sea, vol. 3, Collected Papers (1973).

N. Clark & G. Clark, Governments, Markets and Gas (1984).

T. Daintith ed., The Legal Character of Petroleum Licences: A Comparative Study (1981).

T. Daintith ed., Recent Developments in UK Petroleum Law 1982-1983 (1983).

T.C. Daintith & L. Hancher, Energy Strategy in Europe: The Legal Framework (1986).

T. Daintith & G.D. Willoughby eds., United Kingdom Oil and Gas Law (2nd ed. 1984 (looseleaf)).

A.J. Easson, Tax Law and Policy in the EEC (1980).

Ford Foundation Study Group, Energy: The Next Twenty Years (1979).

L.E. Grayson, National Oil Companies (1981).

R.F. Hayllar & R.M. Rouse, UK Taxation of Offshore Oil and Gas (2nd ed. 1981 (looseleaf)).

G Horwich & D.L. Leo Weimer, Oil Price Shocks, Market Response, and Contingency Planning (1984).

C.W. Howe, Natural Resource Economics: Issues, Analysis and Policy (1979).

International Fiscal Association ed., Taxation of the Extractive Industries (vol. 63a Studies on International Fiscal Law/Cahiers de droit fiscal international, 1978).

D. Ionescu ed., The European Alternatives (1979).

J.P. Kalt, The Economics and Politics of Oil Price Regulation (1981).

W. Keyser & R. Windle, Public Enterprise in the European Economic Community (A Metra Oxford Study, Metra Oxford Consulting Ltd., for the EC Commission, DG IV D4, 1977).

H. Kohler, Welfare and Planning (1966).

R. Lafont, La Révolution Régionaliste (1967).

H.H. Landsberg & J.M. Dukert, High Energy Costs: Uneven, Unfair, Unavoidable? (1981).
D. Lasok & P. Soldatos eds., Les Communautés européennes en fonctionnement/ The European Communities in Action (1981).
S.H. Lay, R. Churchill & M. Nordquist eds., New Directions in the Law of the Sea, vol. 1 Documents (1973).
T.M. Lewis & I.H. McNicoll, North Sea Oil and Scotland's Economic Prospects (1978).
L. Lister, Europe's Coal and Steel Community: An Experiment in Economic Union (1960).
N.J.D. Lucas, Energy and the European Communities (1977).
N.J.D. Lucas, Energy in France: Planning Politics and Policy (1979).
P. McAvoy, Price Formation in Natural Gas Fields (1953).
D. McCloskey, The Applied Theory of Price (1982).
S.L. McDonald, Petroleum Conservation in the United States (1971).
S.L. McDonald, The Leasing of Federal Land for Fossil Fuels Production (1979).
Lord Mackenzie Stuart, The European Communties and the Rule of Law (1977).
E. Mansfield, Microeconomics. Theory and Applications (2d ed. 1975).
J. Mégret, J.-V. Louis, D. Vignes & M. Waelbroeck, Le Droit de la Communauté Economique Européenne (1973).
C.J. Meyers & R.A Posner, Market Transfers of Water Rights: Towards an Improved Market in Water Resources (National Water Commission, Legal Study No. 4, NTIS No. NWC-L-71-009, July 1971).
W. Mönig, Schmitt, Schneider & Schurmann, Konzentration und Wettbewerb in der Energiewirtschaft (1977).
J. Mulholland, Economic Structure and Behavior in the Natural Gas Production Industry (Federal Trade Commission, 1979).
R.H. Nelson, The Making of Federal Coal Policy (1983).
OECD/IEA, Energy Balances of OECD Countries 1960/1974 (1977).
OECD/IEA, Energy Balances of OECD Countries 1970/1982 (1984).
A.M. Okun, Prices and Quantities: A Macro-Economic Analysis (1981).
J.G. Polach, Euratom, Its Background, Issues, Economic Implications (1964).
R. Posner, Economic Analysis of Law (2nd ed. 1977).
D. Prindle, Petroleum Policies and the Texas Railroad Commision (1981).
R. Quadri, R. Monaco, & A. Trabucchi, Trattato istitutivo della Comunità economica europea. Commentario CEE (1965).
R. Quadri, R. Monaco & A.Trabucchi, Trattato istitutivo della Comunità europea del carbone e dell'acciaio. Commentario CECA (1970).
P. Reuter, La Communauté européenne du charbon et de l'acier (1953).
E. Rolph, Nuclear Power and the Public Safety (1979).
M.E. Sanders, The Regulation of Natural Gas: Policy and Politics, 1938-78 (1981).
S.H. Schurr, J. Darmstadter, H. Perry, W. Ramsay & M. Russel, Energy in America's Future: The Choices Before Us (A Study by the Staff of Resources for the Future, 1979).
G.J. Stigler, The Citizen and the State: Essays on Regulation (1975).
A.D. Tarlock ed., Regulation, Federalism, and Interstate Commerce (1981).
L. Tribe, American Constitutional Law (1978).
A. Wenger, Pétrole et gaz naturel en mer du Nord (1971).
D. Wyatt & A. Dashwood, The Substantive Law of the EEC (1980).

Articles

Ackerman & Hassler, *Beyond the New Deal: Coal and the Clean Air Act*, 89 YALE L.J. 1466 (1980).

Allen, *The Euratom Treaty, Chapter VI: New Hope or False Dawn?*, 20 C.M.L. REV. 473 (1983).

Audland, *European Community Energy Strategy and Its Legislative Implications*, 1 J. ENERGY & NAT. RES. L. 9 (1973).

Browde & DuMars, *State Taxation of Natural Resource Extraction and the Commerce Clause: Federalism's Modern Frontier*, 60 ORE. L. REV. 7 (1981).

Bull, *Competition Between Public and Private Enterprises*, in LAW IN THE U.S.A. IN THE BICENTENNIAL ERA 433 (J.N. Hazard & W.J. Wagner eds., supplement to 26 AM. J. COMP. L., 1978).

Campbell, *Principles of Mineral Ownership in the Civil Law and Common Law System*, 31 TULANE L. REV. 303 (1957).

Cappelletti & Golay, *Judicial Review, Transnational and Federal: Its Impact on Integration*, in I/2 INTEGRATION THROUGH LAW 261 (1986).

Cappelletti, Seccombe & Weiler, *Integration Through Law: Europe and the American Federal Experience — A General Introduction*, in I/1 INTEGRATION THROUGH LAW 3 (1986).

Claudy, *The International Energy Agency*, 14 NAT. RES. LAW. 454 (1982).

Coase, *The Problem of Social Cost*, 3 J. LAW & ECON. 1 (1960).

De Bauw, *La Politique énergétique*, in LES COMMUNAUTÉS EUROPÉENNES EN FONCTIONNEMENT/THE EUROPEAN COMMUNITIES IN ACTION 309 (D. Lasok & P. Soldatos eds. 1981).

De Bauw, *Oil Policy Before the Yom Kippur War*, in THE EUROPEAN ALTERNATIVES 79 (G. Ionescu ed. 1979).

Ely, *Legislative and Administrative Motivation in Constitutional Law*, 79 YALE L. J. 1205 (1970).

Evans, *United Kingdom North Sea Oil Policy and EEC Law*, 7 EUR. L. REV. 355 (1982).

Gaja, Hay & Rotunda, *Instruments for Legal Integration in the European Community — A Review*, I/2 INTEGRATION THROUGH LAW 113 (1986).

Garth, *Migrant Workers and Rights of Mobility in the European Community and the United States: A Study of Law, Community and Citizenship in the Welfare State*, in I/3 INTEGRATION THROUGH LAW 85 (1986).

Heller & Pelkmans, *The Federal Economy: Law and Economic Integration and the Positive State — The U.S.A. and Europe Compared in an Economic Perspective*, in I/1 INTEGRATION THROUGH LAW 245 (1986).

Hellerstein, *Constitutional Constraints on State and Local Taxation of Energy-Resources*, 31 NAT. TAX. J. 245 (1978).

Hellerstein, *Constitutional Limitations on State Tax Exportation*, 1982 ABF RESEARCH J. 1.

Hellerstein, *State Taxation and the Supreme Court: Toward a More Unified Approach to Constitutional Adjudication?*, 75 MICH. L. REV. 1426 (1977).

Hellerstein, *State Taxation of Interstate Business and the Supreme Court, 1974 Term:* Standard Pressed Steel *and* Colonial Pipeline, 62 VA. L. REV. 149 (1976).

Jacobs & Karst, *The "Federal" Legal Order: The U.S.A. and Europe Compared — A Juridical Perspective*, in I/1 INTEGRATION THROUGH LAW 169 (1986).

Kitch, *Regulation and the American Common Market*, in REGULATION, FEDERALISM, AND INTERSTATE COMMERCE 10 (A.D. Tarlock ed. 1981).

Kommers & Waelbroeck, *Legal Integration and the Free Movement of Goods: The American and European Experience*, in I/3 INTEGRATION THROUGH LAW 165 (1986).

Krislov, Ehlermann & Weiler, *The Political Organs and the Decision-Making Process in the United States and the European Community*, in I/2 INTEGRATION THROUGH LAW 3 (1986).
Levy, *Relations Between Oil Companies and Consumer State Governments*, 2 J. ENERGY & NAT. RES. L. 9 (1984).
Mead, *The System of Government Subsidies to the Oil Industry*, 10 NAT. RES. J. 113 (1970).
Meyers, *Federal Preemption and State Conservation in* Northern Natural Gas, 77 HARV. L. REV. 689 (1965).
Mitchell, *The Basis of Congressional Energy Policy*, 57 TEX. L. REV. 591 (1979).
Nelson, *Undue Diligence: The Mine-It-or-Lose-It Rule for Federal Coal*, 1983 REGULATION 34 (Jan./Feb. issue).
Note, *Federal Nonreserved Water Rights*, 48 U. CHI. L. REV. 758 (1981).
Note, *Federal Price Control of Natural Gas Sold to Interstate Pipelines*, 59 YALE L.J. 1468 (1950).
Powell, *More Ado About Gross Receipts Taxes*, 60 HARV. L. REV. 501 (1947).
Roeber, *The Formation of North Sea Prices*, in RECENT DEVELOPMENTS IN UK PETROLEUM LAW 1982-83 (T.C. Daintith ed. 1983).
M. Russell, Natural Gas Deregulation: Overview of Policy Issues (Resources for the Future Discussion Paper D-92, Apr. 1982).
Samkalden, *Aantekeningen bij de Ontwerp-Kartelverordening van de Europese Commissie (Art. 87, eerst Lid/EEG)*, 1961 S.E.W. EUROPA 269.
Solow, *Arthur Okun's Last Work (Book Review)*, 65 PUBLIC INTEREST 91 (Fall 1981).
Stein, *The European Community in 1983. A Less Perfect Union?*, 20 C.M.L. REV. 641 (1983).
Stein, *Towards a European Foreign Policy? The European Foreign Affairs System from the Perspective of the United States Constitution*, in I/3 INTEGRATION THROUGH LAW 3 (1986).
Stewart, *Interstate Resource Conflicts: The Role of the Federal Courts*, 6 HARV. ENV. L. REV. 241 (1982).
Tarlock, *Appropriation for Instream Flow Maintenance: A Progress Report on "New" Public Western Water Rights*, 1978 UTAH L. REV. 211 (1978).
Varat, *State "Citizenship" and Interstate Equality*, 48 U. CHI. L. REV. 487 (1981).
Vignes, *The EEC and the Law of the Sea*, in 3 NEW DIRECTIONS IN THE LAW OF THE SEA, COLLECTED PAPERS 335 (R. Churchill, K.R. Simmonds & J. Welch eds. 1973).
Wells & Hellerstein, *The Governmental-Proprietary Distinction in Constitutional Law*, 66 VA. L. REV. 1073 (1980).
Wenger, *La CEE et le plateau continental*, 1971 Revue du Marché Commun 184.
West German Länder to Raise Gas and Oil Royalties to 32 percent, Financial Times, 12 Nov. 1981, at 3.
Williams, *Energy Policy in the Cold Light of Morning*, 61 TEX. L. REV. 571 (1982).
Williams, *Federal Preemption of State Conservation Laws After the Natural Gas Policy Act: A Preliminary Look*, 56 U. COLO. L. REV. 521 (1985).
Williams, *Free Trade in Water Resources:* Sporhase v. Nebraska, *ex rel.* Douglas, 2 S. CT. ECON. REV. 173 (1983).
Williams, *Running Out: The Problem of Exhaustible Resources*, 7 J. LEGAL STUD. 165 (1978).
Williams, *Severance Taxes and Federalism: The Role of the Supreme Court in Preserving a National Common Market for Energy Supplies*, 53 U. COLO. L. REV. 281 (1982).
Williams, *The Requirement of Beneficial Use as a Cause of Waste in Water Resource Development*, 23 NAT. RES. J. 9 (1983).

World Glut Tests the Mettle of BNOC's Oil Traders, Financial Times, 15 Dec. 1982, at 9.

Government Reports

European Community

Comité Intergouvernemental créé par la Conférence de Messine, Rapport des Chefs de Délégation aux Ministres des Affaires Etrangères (Brussels, 21 Apr. 1956) (the *Spaak Report*).

Correspondence Arising Out of the Meeting of the Foreign Ministers of the Governments of Belgium, France, the Federal Republic of Germany, Italy, Luxembourg and the Netherlands held at Messina on June 1-2, 1955, Cmd. 9525 (1955).

EC Commission:

Eighth General Report on the Activities of the European Communities (1974) (1975).

Electricity Prices 1973-1978 (Eurostat, Industry & Services (Blue Series) 1980).

Electricity Prices 1978-1984 (Eurostat, Industry and Services (Blue Covers) 1985).

Gas Prices 1970-1976 (Eurostat, Energy Statistics (Ruby Series) 1977).

Gas Prices 1976-1978 (Eurostat, Energy Statistics (Ruby Series) 1979).

Gas Prices 1978-1984 (Eurostat, Industry & Services, (Blue Covers) 1984).

Ninth Report on Competition Policy (1980).

Seventh General Report on the Activities of the European Communities (1973) (1974).

Sixth Report on Competition Policy (1976) (1977).

Commission Documents:

Memorandum Concerning the Applicability of the Treaty Establishing the European Economic Community to the Continental Shelf, Doc. SEC(70) 3095 final (18 Sept. 1970), *Eng. trans. reprinted in* 10 I.L.M. 202 (1971).

Taxation of Petroleum Products, Doc. COM(81) 511/final (11 Sept. 1981).

Energy Pricing-Policy and Transparency, Doc. COM(81) 539/final (1 Oct. 1981).

The Development of an Energy Strategy for the Community, Doc. COM(81) 540/final (2 Oct. 1981).

Amendment of the Proposal for a Council Decision adopting new provisions relating to Chapter VI, "Supplies", of the Treaty establishing the European Atomic Energy Community, presented by the Commission to the Council in December 1982 (Communication from the Commission to the Council), Doc. COM(84) 606 final/2 (30 Nov. 1984).

ECSC High Authority:

Report on the Problems Posed by the Different Tax Systems Within the Common Market (Brussels, 1953)(the *Tinbergen Committee Report*).

Seventh General Report on the Activities of the Community (1959).

Euratom Commission, Fourth General Report (1961).

United Kingdom

Monopolies and Mergers Commission, Petrol: A Report on the Supply of Petrol in the United Kingdom by Wholesale, Cmd. 7433 (1979).

United States

Congressional Budget Office, The Windfall Profit Tax: A Comparative Analysis of Two Bills (Nov. 1979).

Department of Energy/Energy Information Administration, Energy Programs/ Energy Markets: Overview (DOE/EIA-0201/16, July 1980).

Department of Energy/International Affairs Office of Market Analysis, International Energy Indicators (DOE/IA-0010/14, Aug.-Sept. 1981).

Energy Information Administration, Annual Energy Review 1984.

Federal Energy Regulatory Commission, Notice of Inquiry, Impact of the NGPA on Current and Projected Natural Gas Markets, Docket No. RM82-26-000 (28 Apr. 1982).

Monthly Energy Review (Energy Information Administration, Supt. of Docs., Washington D.C. 1974—).

Natural Gas Annual, 1982, DOE/EIA-0131 (82).

Office of the Secretary of the Treasury, Office of Tax Analysis, Windfall Profit Tax Liabilities and Receipts (25 Jan. 1984, Documentation of 1985 Fiscal Year Budget).

Petroleum Supply Annual, 1982, DOE/EIA-0340 (82/2).

Public Papers of Harry S Truman 1950 (1956).

Quarterly Coal Report, Oct./Dec. 1983, DOE/EIA-0121 (84/4Q).

State Energy Data Report, Consumption Estimates, 1960-82, DOE/EIA-0214 (84).

Statistical Data of the Uranium Industry, Jan. 1, 1983, GJO-100 (83).

Index

Index of Enterprises

Index of Authors and Names

Integration Through Law

Europe and the American Federal Experience

A Series under the General Editorship of

Mauro Cappelletti · Monica Seccombe · Joseph Weiler

Part One

Volume 1

Methods, Tools and Institutions (3 Books)

edited by Mauro Cappelletti, Monica Seccombe & Joseph Weiler

Book 1: A Political, Legal and Economic Overview (1986, XC & 616 pp.)
Book 2: Political Organs, Integration Techniques and Judicial Process (1986, XXX & 351 pp.)
Book 3: Forces and Potential for a European Identity (1986, XXX & 405 pp.)

Part Two

Integration Policies in Selected Areas

Volumes 2—6

Volume 2: Environmental Protection Policy
by
Eckard Rehbinder & Richard Stewart (1985, XXIV & 351 pp.)

Volume 3: Consumer Law, Common Markets and Federalism in Europe and the United States
by
Thierry Bourgoignie & David Trubek (with Louise Trubek & Denis Stingl) (1987, XXIV & 272 pp.)

Volume 4: Legal Harmonization and the Business Enterprise: Corporate Law and Capital Market Harmonization Policy in Europe and the U.S.A.
by
Richard Buxbaum & Klaus J. Hopt (forthcoming)

Volume 5: The Legal Integration of Energy Markets
by
Terence Daintith & Stephen Williams (1987, XXVIII & 176 pp.)

Volume 6: Regionalism and Federalism: The Challenge of Regions in National and Transnational Polities
by
Yves Mény & Bruno De Witte (forthcoming)

European University Institute. Series A [Law]: 2

Bei Fragen zur Produktsicherheit wenden Sie sich bitte an:
If you have any questions regarding product safety,
please contact:

Walter de Gruyter GmbH
Genthiner Straße 13
10785 Berlin
productsafety@degruyterbrill.com